R. H. F. Hunter

The Fallopian Tubes

Their Role in Fertility and Infertility

With 55 Figures, 2 in Colour

Springer-Verlag
Berlin Heidelberg New York
London Paris Tokyo

Dr. RONALD H. F. HUNTER
Faculty of Science
University of Edinburgh
West Mains Road
Edinburgh EH9 3JG
Scotland, Great Britain

Present address

Faculty of Veterinary Medicine (CRRA)
University of Montreal
CP 5000, Saint Hyacinthe
Quebec, Canada J2S 7C6

ISBN 3-540-18436-8 Springer-Verlag Berlin Heidelberg New York
ISBN 0-387-18436-8 Springer-Verlag New York Berlin Heidelberg

Library of Congress Cataloging-in-Publication Data. Hunter, R.H.F. The fallopian tubes: their role in fertility and infertility / R.H.F. Hunter.   p.   cm. Includes index. ISBN 0-387-18436-8 (U.S.) 1. Fallopian tubes. 2. Fallopian tubes – Diseases. 3. Fertilization (Biology). 4. Fertilization in vitro, Human. I. Title. [DNLM: 1. Fallopian Tubes – physiology. WP 300 H9455f]. QP265.H86   1988.   612'.62–dc19.   DNLM/DLC for Library of Congress. 87-36951 CIP.

© Springer-Verlag Berlin Heidelberg 1988
Printed in Germany

Typesetting: K+V Fotosatz GmbH, Beerfelden
Printing and Binding: Konrad Triltsch, Graphischer Betrieb, Würzburg
2131/3130-543210

Preface

This monograph has been written in the hope that it will prove of value to medical students and clinicians, to Honours undergraduates in appropriate branches of the natural sciences, and to reproductive biologists in general. It would be pleasing if the text also caught the attention of veterinary undergraduates, since there is much information bearing on reproduction in domestic animals. First and foremost, however, the intended audience is a medical one, for scientific studies of human reproduction have been catalyzed by the intense interest in procedures of fertilization in vitro. Some would judge that this very activity has narrowed our view of physiological events occurring within the Fallopian tubes. The present work may therefore serve as a useful counterbalance to the overwhelming series of publications on procedures of in vitro fertilization, and offer opportunities to those in the clinical field for extending their knowledge of the scientific background to much of the current work.

The writing did not commence with such a purpose. Since his first years as a research student, the author has been especially interested in the process of fertilization and the beginning of embryonic development. These interests expanded with a series of surgical and pharmacological interventions used to modify the function of the Fallopian tubes in experimental animals. By the early 1970s, such studies were attracting the attention of medical audiences and by the mid 1970s, invitations to lecture at home and abroad were a pleasingly frequent occurrence. On the basis of this activity, the author published a detailed review in the European Journal of Obstetrics, Gynaecology and Reproductive Biology (1977) entitled *Function and malfunction of the Fallopian tubes in relation to gametes, embryos and hormones*. This was well received, although it remains uncertain whether the text or the coloured photographs proved the greater attraction! An up-dated version of the review was written in 1980 (but not published until 1982), under the title of *Anatomy and physiology of the Fallopian tube*. In part, therefore, the origins of the present volume must lie in these earlier attempts. The writings of others have clearly inspired the author, and a much-prized possession is a copy of *The Mammalian Oviduct* edited by Hafez and Blandau (1969). Other sources of inspiration are mentioned in Chapter I.

A monograph is, of course, one man's view in which he can indulge at least some of his own prejudices. Readers will all too readily appreciate the limitations to this situation, but one of the advantages is that − even in 1987 − an author can still express his own thoughts in his own words rather than accepting the uniform vocabulary imposed by most scientific editors. Within this welcome freedom, the chapters have been written to stand individually, although with a reasonable amount of cross-reference. This approach has involved some overlap of specific points or themes, but nowhere is this extensive. As to a policy on references, the intention has been to be selective rather than exhaustive; it is nonetheless hoped that the citations to key works and reviews will provide access to the wider body of literature.

If the present work leads to an appreciation of the subtle and dynamic interactions occurring locally between the embryo, the Fallopian tube and the neighbouring ovary, and perhaps prompts further experiments, then its purpose will have been well served. The fact of a physiological conversation between the newly-fertilized egg, the enveloping duct system, and the adjacent gonad should not cause surprise in well-read audiences of the 1980s. However, the nuances of that conversation will undoubtedly keep us entertained for many years to come, and indeed the precocity of the zygote may well take some unawares.

Edinburgh, June 1987   R. H. F. HUNTER

Acknowledgements

In most academic endeavours, one remains conscious of the impact of one's teachers and in this respect I have been particularly fortunate. My warmest thanks therefore go straight away to Dr. E. J. C. Polge, FRS, Prof. T. R. R. Mann, FRS and Mr. L. E. A. Rowson, FRS in Cambridge for giving me my first opportunities in research and for offering advice and support since 1961. During three post-doctoral years in the United States, Prof. P. J. Dziuk and the late Prof. A. V. Nalbandov of the University of Illinois made a lasting impression, and Dr. M. C. Chang of the Worcester Foundation for Experimental Biology was also an inspiring teacher. The same was true of Prof. C. Thibault of the University of Paris and Station Centrale de Physiologie Animale, Jouy-en-Josas. To each of these distinguished physiologists, I should like to record my debt of gratitude.

This book could not have been written without the enthusiastic support of various friends and colleagues in Edinburgh. Mrs. Pat Gallie devoted an enormous amount of time, care and patience to typing drafts of the manuscript, and then the final version, as did Mrs. Frances Anderson for the lists of references. Mr. Robert Nichol gave endless assistance, primarily in tracking down a host of references and obscure volumes, but also with his artistic endeavours and as a skilled anaesthetist in many of the surgical experiments referred to in the text. Dr. T. A. Bramley of the Department of Obstetrics and Gynaecology provided three of the line drawings, and the photographic plates were skilfully prepared by Mr. Gordon Finnie and Mr. David Stewart Robinson. Miss Ruth Johnson was especially helpful in obtaining photocopies of rare material.

Friends in Edinburgh who kindly cast an eye over an early draft of the manuscript were Dr. Charlotte Chalmers and Fiona Rhodes, MRCVS, together with Drs. A. A. Macdonald, C. A. Price and L. C. Smith. Dr. Brian Cook of the University of Glasgow generously provided detailed comments on a draft of Chapter III. To all of these people, I express my best thanks for time and trouble taken on my behalf.

Colleagues in the UK and beyond who donated reprints, preprints or specific information include: Drs. J. D. Biggers, K. P. Bland, O. Bomsel-Helmreich, M. C. Chang, J. Cohen, B. Cook, N. Einer-Jensen, J. E. Fléchon, M. J. K. Harper, J. P. Hearn, R. B. Heap, H. J. Leese, M. C. Levasseur, A. A. Macdonald, J. H. Marston, Y. Menezo, D. Mortimer, C. O'Neill, M. E. Perotti, E. J. C. Polge, N. L. Poyser, C. A. Price, R. F. Seamark, G. Simm, J. P. Simons, L. C. Smith, D. Szöllösi, C. Thibault, S. Torres, C. Wathes, R. Webb, I. Wilmut and R. Yanagimachi. Assistance with illustrative material or permission to copy figures and diagrams was offered by Mr. P. K. Escreet of Glasgow University Library, Mrs. J. Currie of Edinburgh University Library, the Bibliothèque Nationale, Paris, and: Drs. J. M. Bedford, A. R. Bellve, R. J. Blandau, O. Bomsel-Helmreich, J. E. Fléchon, V. Gomel, G. S. Greenwald, M. J. K. Harper, M. Johnson, C. Nancarrow, C. O'Neill, C. J. Pauerstein, B. J. Restall, D. Szöllösi, C. Thibault, A. Trounson and R. Yanagimachi. Gratitude is also expressed to the Editors of the following journals for permission to reproduce graphical and/or tabular material: Anatomical Record, Annals of the New York Academy of Sciences, European Journal of Obstetrics, Gynaecology and Reproductive Biology, Fertility and Sterility, International Journal of Fertility, Journal of Anatomy, Journal of Endocrinology, Journal of Experimental Zoology, Journal of Reproduction and Fertility, and Reproduction, Nutrition, Développement; and to the following publishing houses: Academic Press, Cambridge University Press, Excerpta Medica, Karger, Lea and Febiger and Plenum Press.

Finally, I should like to record my warmest thanks to Dr. D. Czeschlik of Springer-Verlag, Heidelberg, for flexibility and encouragement during an unforeseen delay; Mr. P. J. Abernethy, FRCSE and Miss A. Robertson of the Western General Hospital, Edinburgh, for specific assistance and guidance through a somewhat trying spell; and most of all to my wife for continuous help with preparation and checking of the manuscript, and for tolerating the peculiar habits associated with writing and research.

Contents

I   *Discovery of the Fallopian Tubes and Subsequent
    Historical Landmarks* 1

Introduction 1
Ancient Interpretations 1
Sixteenth and Seventeenth Century Views 2
Eighteenth and Nineteenth Century Advances 4
Twentieth Century Studies: Experimental Trends 7
References 9

II   *Development of the Fallopian Tubes and
     Their Functional Anatomy* 12

Introduction 12
Embryonic Derivation of the Fallopian Tube 14
Vascular Anatomy of the Tube 17
The Lymphatic System 19
Musculature and Innervation 20
Mucosa: Cilia and Secretory Cells 24
References 27

III *Fallopian Tube Fluid: The Physiological Medium for
     Fertilization and Early Embryonic Development* 30

Introduction 30
Dynamics of Tubal Fluid Accumulation 31
Direction of Fluid Flow 33
Composition of Tubal Fluids 34
Regional and Microenvironments 43
Hormones Detected in Tubal Fluids 45
References 48

IV   *Transport of Gametes, Selection of Spermatozoa, and
     Gamete Lifespans in the Female Tract* 53

Introduction 53
Shedding and Initial Transport of Eggs 53

Deposition and Transport of Spermatozoa   58
Pre- and Peri-Ovulatory Sperm Distribution Within the
    Isthmus   66
Intra-Peritoneal Insemination   70
Sperm Selection in the Female Tract   71
Gamete Lifespans in the Female Tract   72
References   74

V   Denudation of Eggs, Capacitation of Spermatozoa,
       and Fertilization − Normal and Abnormal   81

Introduction and Chemotaxis   81
Denudation and Final Maturation of Eggs   82
Fertilizin-Antifertilizin Reactions at the Egg Surface   84
Binding Sites for Spermatozoa on the Zona Pellucida   85
Capacitation of Spermatozoa, the Acrosome Reaction
    and Hyperactivation   88
Penetration of the Zona and Fusion of Gametes   96
Activation of the Egg and Block to Polyspermy   97
Synchrony of Penetration and Cleavage   99
Abnormalities of Fertilization   101
Fate of Redundant Gametes   102
References   103

VI   Development of the Embryo and Influences on the
       Maternal System   109

Introduction   109
Developmental Stages in the Fallopian Tube   110
Nutritional Aspects of Embryonic Development   112
Early Pregnancy Factors   115
Experimental Modification of Tubal Development   120
Organization of the Embryo: Some First Steps   120
References   122

VII   Transport of Embryos to the Uterus: Normal and
       Abnormal Timing   127

Introduction   127
Progression of Embryos Within the Fallopian Tube   128
Mechanisms Regulating Passage of Embryos   130
Pharmacological Disturbance of Egg Transport   133
Delayed Transport in the Aetiology of Tubal Pregnancy . .   135
References   136

VIII *Malfunction of the Fallopian Tubes: Spontaneous Conditions and Surgical Studies* 139

Introduction 139
Anomalies of the Fallopian Tubes 140
Ectopic (Tubal) Pregnancy 141
Intra-Uterine Fertilization and Estes' Operation 145
Tubal Resection and Reconstructive Surgery............ 149
Microsurgery and Subsequent Pregnancy 153
References 156

IX *In Vitro Fertilization, Manipulation of Eggs and Embryos, and Subsequent Transplantation* 160

Introduction 160
Historical Aspects 161
Experimental Conditions for In Vitro Fertilization 163
Clinical Approach to In Vitro Fertilization 165
Embryo Culture and Transplantation into the Uterus 168
Alternative Approaches to Infertility 171
Sperm-Penetration of Zona-Free Oocytes 172
Other Cellular and Nuclear Manipulations 174
References 177

X *Are the Fallopian Tubes Essential? A Biological Perspective* 183

Subject Index 185

Abbreviations, Units and Adjectives

EPF	Early pregnancy factor
FSH	Follicle stimulating hormone
GIFT	Gamete intra-Fallopian transfer
GnRH	Gonadotrophin releasing hormone
HCG	Human chorionic gonadotrophin
HMG	Human menopausal gonadotrophin
IVF	In vitro fertilization
LH	Luteinizing hormone
PAS	Periodic acid-Schiff
$PGF_{2\alpha}$	Prostaglandin $F_{2\alpha}$
PMSG	Pregnant mares' serum gonadotrophin

mg	milligram (10^{-3} g)
µg	microgram (10^{-6} g)
ng	nanogram (10^{-9} g)
pg	pikogram (10^{-12} g)

i.u.	international units
mOsmols	milliosmols
M_r	relative molecular mobility (molecular weight)

Distal	uterine end of Fallopian tube (i.e. isthmus)
Proximal	ovarian end of Fallopian tube (i.e. ampulla)

CHAPTER I

Discovery of the Fallopian Tubes and Subsequent Historical Landmarks

CONTENTS

Introduction 1
Ancient Interpretations 1
Sixteenth and Seventeenth Century Views 2
Eighteenth and Nineteenth Century Advances 4
Twentieth Century Studies: Experimental Trends 7
References 9

Introduction

In the context of the reproductive system, the names of Fallopius and de Graaf are widely known to students working in this now distinctive field of physiology. However, it is historians rather than physiologists who are perhaps best placed to appreciate the remarkable good fortune of Fallopius (Gabriele Fallopio 1523–1562) in the sphere of medical discovery. Almost a contemporary of Andreas Vesalius at Padua, he was at first a pupil of this distinguished anatomist and yet it was to be the name of Fallopius — and not that of his master — which would lend itself to quite specialised portions of the female genital tract. How did this situation arise? The history of earlier endeavours appears in part to be one of misinterpretation coupled with a too-willing acceptance of the dogma handed down the ages. Instead of reviewing this saga in detail, a task which itself could lead to the writing of a weighty volume, only a few highlights and personalities are touched upon in the following pages. The objective has been to pave the way for the more substantial contents of subsequent chapters.

Ancient Interpretations

The story commences in antiquity, possibly with brief mention of the tubes in Hindu medical writings some 1000–800 B.C. However, it was the foundations laid by investigations in ancient Greece and Egypt that were to remain prominent in the scholarly world for many hundreds of years. Because the rôle of the ovaries in the reproductive process was far from being unravelled, attention was focussed largely on the anatomy of the uterus, the organ known to contain the developing foetus. Aristotle (384–322 B.C.) proposed that the embryo was generated by means of an intermingling of the male semen with the uterine blood or menstrual fluids, the latter being viewed as female semen. Although he noted structures apparently linking the ovaries to the uterus, Aristotle did not draw specific attention to the Fallopian tubes, except by remarking upon their convolutions. Herophilus, a widely-renowned 3rd century B.C. anatomist of Alexandria, also described the paired ducts in his major treatise on midwifery, but presumed that the ducts transmitted semen – female semen – from the ovaries to the urinary bladder. Soranus of Ephesus (circa A.D. 100), a Greek physician who practised medicine in Rome and was a reputedly prolific writer, perpetuated this interpretation in his classic work on gynaecology, with the added misunderstanding that "the female seed seems not to be drawn upon in generation, since it is excreted externally" (Temkin 1956).

Despite the gradual decline of the scientific centre in Egypt, Galen (A.D. 130–200), a Greek physician and biologist who similarly progressed to Rome, crossed the Mediterranean to study at Alexandria from A.D. 152–157. As might be supposed, he was there much influenced by the traditions of the great medical school and the writings of Herophilus. Even so, he was not constrained by prevailing views and correctly described the paired ducts as leading to the uterus and not to the bladder. In fact, based on observations in domestic animals with a bicornuate uterus, he described the ducts as terminating in the tips of the uterine horns. His mistake was to consider these ducts as the female counterpart of the male seminal ducts, carrying "female semen" filtered by the ovaries from the bloodstream down into the uterus.

What may have been thought during the Dark Ages (and strictly-speaking also the Middle Ages) is scarcely relevant here, save to remark that scientific interpretations of the reproductive organs did not take any steps forward. The doctrines of Galen remained prominent for almost fourteen centuries after his death – a mark of his stature in medical history and the extensive influence of his writing. The wonderful drawings of Leonardo da Vinci (1452–1519) took shape towards the end of this period but, despite depicting the uterus and adjoining structures, they neither dwelt on the nature of the duct system nor were they published systematically for the benefit of the scientific world until the late nineteenth century.

Sixteenth and Seventeenth Century Views

In 1543, Andreas Vesalius published his masterpiece entitled *De Humani Corporis Fabrica* (the seven volumes usually being known as the *Fabrica*), with its extensive account of the anatomy of the generative organs. Whilst there is little doubt that he carefully noted the disposition of the Fallopian tubes and indeed depicted them in Book V of the *Fabrica* (Fig. I.1), Vesalius failed to clarify their specific function. Instead, he considered their relationship with the ovaries was analogous to that of the male ducts with the testes, and therefore that they were "semen-conveying" vessels; in effect, he was echoing the views of Galen, and illustrated the ducts as coiled round the "female testes". Of course, the uterus was still regarded as the origin of the embryo: no notion was yet abroad for the ex-

2

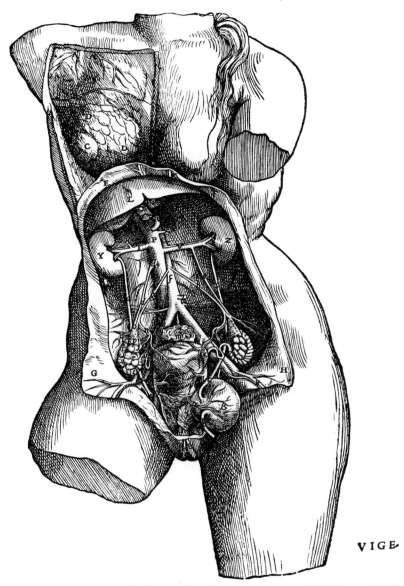

Fig. I.l. No essay dealing with the female reproductive system can overlook this wonderful illustration from Vesalius (1543), which appeals on aesthetic, anatomical and historical grounds.
(Courtesy of Glasgow University Library)

istence of male and female gametes — let alone for the ovary as a source of the oocyte — so it is difficult to be unduly critical of Vesalius. Whether Fallopius genuinely brought a more talented mind to bear on the problems of generation would seem questionable, but his observation that the duct was a trumpet-shaped tube, open to the abdomen and leading into the lumen of the uterus was undoubtedly a step forward. In his *Observationes anatomicae* (1561), Fallopius wrote the definitive description and thereby lent his name:

This seminal duct (*meatus seminarius*) originates from the cornua uteri; it is thin, very narrow, of white colour and looks like a nerve. After a short distance it begins to broaden and to coil like a tendril (*capreolus*), winding in folds almost up to the end. There, having become very broad, it shows an *extremitas* of nature of skin and colour of flesh, the utmost end being very ragged and crushed, like the fringe of worn out clothes. Further, it has a great hole which is held closed by the fimbriae which lap over each other. However, if they spread out and dilate, they create a kind of opening which looks like the flaring bell of a brazen tube. Because the course of the seminal duct, from its origin up to its end, resembles the shape of this classical instrument — anyhow, whether the curves are existing or not — I named it *tuba uteri*. These uterine tubes are alike not only in men, but also in the cadavers of sheeps and cows, and all the other animals which I dissected. (Herrlinger and Feiner 1964).

Notable observations and hypotheses in the succeeding century can be summarised briefly; more detailed comment is to be found in the excellent review by Bodemer (1969). Also in Padua, a pupil of Fallopius called Fabricius (1537 – 1619) interpreted the tube or duct as an organ of secretion, having rendered a reasonably accurate account of its function in the formation of chicken eggs. Although Harvey (1578 – 1657), a sometime student of Fabricius, failed to make a major statement on the Fallopian tubes, he did conclude that the existence of "female semen" was a myth; this stemmed from his examination of the uterus in various mammals after mating, and especially his studies of deer, in which no trace of conception was found for a protracted interval. Nonetheless, Harvey (1651), was inspired to propose that "All living things come from eggs" (Fig. I.2).

Van Horne (1668) thought that the Fallopian tubes might function to transmit vesicles (i.e. follicles, but he called them eggs) to the uterus, and de Graaf (1672), initially working with avian ovaries (Fig. I.3), identified the tertiary follicles that were later to take his name in mammalian ovaries. He maintained that the eggs originated here, but the transport function of the Fallopian tubes was not yet generally appreciated: Harvey's notion of conception — of an ovum formed in utero — still prevailed. By 1672, however, de Graaf had traced the would-be passage of eggs through the tubes to the uterus, believing that the follicles were eggs and noting their disappearance after copulation in the rabbit (an induced ovulator) followed by the appearance of blastocysts in the uterus a few days later. But, as he himself realized, the problem of the diameter of follicles vis à vis that of the Fallopian tubes could not be reconciled with passage of eggs, unless it was in fact the contents of the follicles that entered the tubes.

Eighteenth and Nineteenth Century Advances

With improving models of microscope becoming available, de Graaf's observations were eventually endorsed by William Cruickshank (1797), who isolated and

Gulielmus Harveus
de
Generatione Animalium.

Fig. I.2. The frontispiece of William Harvey's (1651) book, showing Zeus liberating living beings from an egg. The caption indicates that 'All living things come from eggs'. (Courtesy of Edinburgh University Library)

identified rabbit eggs from the tube; the mucin layer that develops around the eggs of this species undoubtedly aided his task. By 1827, von Baer had discovered the origin of the mammalian egg, and had attempted to remove the confusion existing between eggs and follicles. This was achieved by dissecting open Graafian follicles and examining the liberated contents within their cumulus masses. Only after the realization that the sperm enters the egg at fertilization (Barry 1843; Bischoff 1854), and then the examination of these events in a systematic manner (Van Beneden 1875), did correspondingly more detailed information become available concerning the Fallopian tubes — this was not least because histological sections of the tubes were used in many of the classical studies of fertilization and early embryonic development (e.g. Sobotta 1895; Lams 1913).

Physiological observations had tentatively taken to the stage, with Blundell (1819) referring to the muscular movements of the tube and uterus. Bischoff (1842) suggested that contractions of the tube and its mesenteries, together with ciliary activity, were necessary for the eggs to pass to the uterus, and Thiry (1862) proposed that ciliary activity was responsible for the movement of eggs from the ovaries actually into the tubes. Experiments bearing on the rate of passage of eggs along the tubes were performed by Pinner (1880), who noted that particulate matter — such as an India ink suspension introduced to the tubes at the ovarian end — could be transported to the vagina within hours.

Twentieth Century Studies: Experimental Trends

Table I.1 presents some landmarks in the study of Fallopian tube function from 1925 onwards, and illustrates in the left-hand column the range of the experimental approaches.

Not listed, however, are the many distinguished works on the Fallopian tubes and embryos emanating largely from the Johns Hopkins Medical School and the Carnegie Institution of Washington in the first 30 years of this century, such as those of Andersen (1927a, b, 1928), Corner (1921, 1923), Heuser (1927), Heuser and Streeter (1929), Seckinger (1923), Seckinger and Snyder (1926), Snyder (1923, 1924), and Wislocki and Guttmacher (1924). Although much of this research activity concerned human material, studies on genital tracts of the domestic pig also featured prominently. Many of the publications were largely of an anatomical nature, but physiological aspects of tubal function were reported in 1923 after using the kymographic technique to record the contractile activity of strips of tube and/or uterus obtained at different stages of the oestrous cycle or pregnancy (Seckinger 1923).

The subsequent landmarks summarized in Table I.1 include direct observations of tubal contractility in vivo, measurement of rates of egg transport under different endocrine conditions and, more recently, the application of biochemical and pharmacological techniques to the study of tubal function. The advent of

---

Fig. I.3. The ovary and duct system of a chicken as portrayed by de Graaf (1672). The ovarian follicles are much larger than the corresponding mammalian structures, for they become distended with yolk. (Courtesy of Edinburgh University Library)

Table I.1. Some landmarks in the study of Fallopian tube function from 1925 onwards. (Adapted from Hunter 1977)

Nature of study	Species	Reference
Muscular contractions of tube observed via abdominal window technique	Rabbit	Westman (1926)
Egg transport observed by		
a) dissection technique,	Rabbit	Greenwald (1961)
b) vital staining of cumulus mass, or	Rabbit	Harper 1961)
c) use of artificial eggs	Sheep	Wintenberger-Torres (1961)
Biochemical composition of tubal fluids collected by cannulation	Rabbit	Hamner and Williams (1965)
	Sheep	Perkins et al. (1965)
	Monkey	Mastroianni et al. (1961)
Pharmacology and neuro-pharmacology of tubal isthmus	Rabbit	Brundin (1964)
	Sheep	Holst et al. (1970)
	Human	Sjöberg (1967)
Electron microscopy of cilia and secretory cells	Mouse	Nilsson and Reinius (1969)
	Monkey	Brenner (1969a, b)
Scanning electron microscopy of tubal epithelium	Rabbit	Rumery and Eddy (1974)
	Cow, pig, goat	Stalheim et al. (1975)
	Human	Patek (1974)
Involvement of prostaglandins in tubal function	Rabbit, sheep	Horton et al. (1965)
	Rabbit	Spilman and Harper (1973)
	Human	Coutinho and Maia (1971)
Extensive clinical and surgical work	Human	Estes and Heitmeyer (1934)
		Rubin (1947)
		Palmer (1960)
Luminal fluid analysis in situ	Mouse	Roblero et al. (1976)
	Human	Borland et al. (1980)
Microsurgical studies	Rabbit	Pauerstein (1978)
	Human	Winston (1980)
	Human	Gomel (1980)
Biochemical potential of tube perfused in vitro	Rabbit	Leese (1983)
		Leese and Gray (1985)
Early pregnancy factors	Mouse	Morton et al. (1976)
	Sheep, cow	Nancarrow et al. (1981)
	Mouse	O'Neill (1985)
	Human	O'Neill et al. (1985)

scanning electron microscopy has stimulated further interest in functional anatomy, and has aided interpretation of the interrelationship between gametes, zygotes and the tubal epithelium. Throughout these various phases of interest, there have been extensive clinical and surgical studies of human Fallopian tubes, embracing such topics as Rubin's (1920) transuterine insufflation test for tubal patency, the development of a technique for transplanting ovarian tissue into the uterus (Estes 1924), and numerous attempts at tubal resection and anastomosis (Green-Armytage 1959; Palmer 1960). Microsurgery has been in vogue in some quarters for the last 10 years or so, but there remain voices critical of these developments (Chap. VIII). In any event, advances in the techniques for in vitro fertilization of the human egg and transplantation of the embryo have helped to by-pass the problem of obstructed Fallopian tubes (Chap. IX), and removed the need for abdominal surgery.

Much of the experimental work in laboratory and domestic animals was reviewed in *The Mammalian Oviduct* edited by Hafez and Blandau (1969) and in *The Oviduct and its Functions* edited by Johnson and Foley (1974); an authoritative survey of human studies has also been published (Woodruff and Pauerstein 1969). Other treatments dealing with the function of the Fallopian tubes include those of Thibault (1972), Pauerstein (1975), and a valuable coverage in the *Handbook of Physiology* (Greep 1973). The WHO symposium volume entitled *Ovum Transport and Fertility Regulation* (Harper et al. 1976) also assembles a wealth of relevant information on laboratory animals and primates. The reviews of Hunter (1977, 1982) have already been mentioned in the Preface. More specialised treatments are referred to in the individual chapters that follow.

REFERENCES

Andersen DH (1927a) Lymphatics of the Fallopian tube of the sow. Contrib Embryol Carneg Instn 19:135–148

Andersen DH (1927b) The rate of passage of the mammalian ovum through various portions of the Fallopian tube. Am J Physiol 82:557–569

Andersen DH (1928) Comparative anatomy of the tubo-uterine junction. Histology and physiology in the sow. Am J Anat 42:255–305

von Baer KE (1827) De ovi mammalium et hominis genesi. Leipzig

Barry M (1843) Spermatozoa observed within the mammiferous ovum. Phil Trans Roy Soc B 133:33

Bischoff TLW (1842) Entwicklungsgeschichte des Kanincheneies. Braunschweig

Bischoff TLW (1854) Entwicklungsgeschichte des Rehes. Gießen

Blundell J (1819) Experiments on a few controverted points respecting the physiology of generation. Med Chir Soc Trans 10:245–272

Bodemer CW (1969) History of the mammalian oviduct. In: Hafez ESE, Blandau RJ (eds) The mammalian oviduct. University of Chicago Press, Chicago, pp 3–26

Borland RM, Biggers JD, Lechene CP, Taymor ML (1980) Elemental composition of fluid in the human Fallopian tube. J Reprod Fertil 58:479–482

Brenner R (1969a) The biology of oviductal cilia. In: Hafez ESE, Blandau RJ (eds) The mammalian oviduct. University of Chicago Press, Chicago, pp 203–229

Brenner R (1969b) Renewal of oviduct cilia during the menstrual cycle in the Rhesus monkey. Fertil Steril 20:599–611

Brundin J (1964) The distribution of noradrenaline and adrenaline in the Fallopian tube of the rabbit. Acta Physiol Scand 62:156–159

Corner GW (1921) Abnormalities of the mammalian embryo occurring before implantation. Contrib Embryol Carneg Instn 13:61–66

Corner GW (1923) Cyclic variation in uterine and tubal contraction waves. Am J Anat 32:345–351

Coutinho EM, Maia H (1971) The contractile response of the human uterus, Fallopian tubes and ovary to prostaglandins in vivo. Fertil Steril 22:539–543

Cruickshank W (1797) Experiments in which, on the third day after impregnation, the ova of rabbits were found in the Fallopian tubes; and on the fourth day after impregnation in the uterus itself; with the first appearance of the foetus. Phil Trans Roy Soc 87:197–214

De Graaf R (1672) De mulierum organis generationi inservientibus. Leyden

Estes WL Jr (1924) Ovarian implantation. Surg Gynecol Obstet 38:394–398

Estes WL Jr, Heitmeyer PL (1934) Pregnancy following ovarian implantation. Am J Surg 24:563–580

Fallopius G (1561) Observationes anatomicae. Venice

Gomel V (1980) Microsurgical reversal of female sterilization: a reappraisal. Fertil Steril 33:587–597

Green-Armytage VB (1959) Recent advances in the surgery of infertility. J Obstet Gynaecol Brit Emp 66:32–39

Greenwald GS (1961) A study of the transport of ova through the rabbit oviduct. Fertil Steril 12:80–95

Greep RO (ed) (1973) Handbook of physiology. Section 7 Endocrinology II. Female Reproductive System Part II. American Physiological Society, Washington DC

Hafez ESE, Blandau RJ (eds) (1969) The mammalian oviduct. University of Chicago Press, Chicago

Hamner CE, Williams WL (1965) Composition of rabbit oviduct secretions. Fertil Steril 16:170–176

Harper MJK (1961) The mechanisms involved in the movement of newly ovulated eggs through the ampulla of the rabbit Fallopian tube. J Reprod Fertil 2:522–524

Harper MJK, Pauerstein CJ, Adams CE, Coutinho EM, Croxatto HB, Paton DM (1976) Symposium on ovum transport and fertility regulation. Scriptor, Copenhagen

Harvey W (1651) Exercitationes de generatione animalium. London

Herrlinger R, Feiner E (1964) Why did Vesalius not discover the Fallopian tubes? Med Hist 8:335–341

Heuser CH (1927) A study of the implantation of the ovum of the pig. Contrib Embryol Carneg Instn 19:229–243

Heuser CH, Streeter GL (1929) Early stages in the development of pig embryos. Contrib Embryol Carneg Instn 20:1–30

Holst PJ, Cox RI, Braden AWH (1970) The distribution of noradrenaline in the sheep oviduct. Aust J Exp Biol Med Sci 48:563–565

Horton EW, Main IHM, Thompson CJ (1965) Effects of prostaglandins on the oviduct, studied in rabbits and ewes. J Physiol 180:514–528

Hunter RHF (1977) Function and malfunction of the Fallopian tubes in relation to gametes, embryos and hormones. Europ J Obstet Gynecol Reprod Biol 7:267–283

Hunter RHF (1982) Anatomy and physiology of the Fallopian tube. In: Chamberlain G, Winston R (eds) Tubal infertility: diagnosis and treatment. Blackwell Sci Publ, Oxford, pp 1–29

Johnson AD, Foley CW (eds) (1974) The oviduct and its functions. Academic Press, New York

Lams H (1913) Étude de l'oeuf de cobaye aux premiers stades de l'embryogenèse. Arch Biol Paris 28:229–323

Leese HJ (1983) Studies on the movement of glucose, pyruvate and lactate into the ampulla and isthmus of the rabbit oviduct. Q Jl Exp Physiol 68:89–96

Leese HJ, Gray SM (1985) Vascular perfusion: a novel means of studying oviduct function. Am J Physiol 248:E624–E632

Mastroianni L Jr, Shah U, Abdul-Karim R (1961) Prolonged volumetric collection of oviduct fluid in Rhesus monkey. Fertil Steril 12:417–424

Morton H, Hegh V, Clunie GJA (1976) Studies of the rosette inhibition test in pregnant mice: evidence of immunosuppression. Proc R Soc B 193:413–419

Nancarrow CD, Wallace ALC, Grewal AS (1981) The early pregnancy factor of sheep and cattle. J Reprod Fertil Suppl 30:191–199

Nilsson O, Reinius S (1969) Light and electron microscopic structure of the oviduct. In: Hafez ESE, Blandau RJ (eds) The mammalian oviduct. University of Chicago Press, Chicago, pp 57–83

O'Neill C (1985) Thrombocytopenia is an initial maternal response to fertilisation in mice. J Reprod Fertil 73:559–566

O'Neill C, Pike IL, Porter RN, Gidley-Baird A, Sinosich MJ, Saunders DM (1985) Maternal recognition of pregnancy prior to implantation: methods for monitoring embryonic viability in vitro and in vivo. Annls NY Acad Sci 442:429–439

Palmer R (1960) Salpingostomy – a critical study of 396 personal cases operated upon without polythene tubing. Proc R Soc Med 53:357–359

Patek E (1974) The epithelium of the human Fallopian tube. Acta Obstet Gynecol Scand 53 Suppl 31:1–28

Pauerstein CJ (1975) Clinical implications of oviductal physiology and biochemistry. Gynecol Invest 6:253–264

Pauerstein CJ (1978) From Fallopius to fantasy. Fertil Steril 30:133–140

Perkins JL, Goode L, Wilder WA, Henson DB (1965) Collection of secretions from the oviduct and uterus of the ewe. J Anim Sci 24:383–387

Pinner O (1880) Über den Übertritt des Eies aus dem Ovarium in die Tube beim Säugetier. Arch Physiol 241

Roblero L, Biggers JD, Lechene CP (1976) Electron probe analysis of the elemental microenvironment of oviducal mouse embryos. J Reprod Fertil 46:431–434

Rubin IC (1920) The nonoperative determination of patency of Fallopian tubes. J Am Med Assn 75:661–666

Rubin IC (1947) Intra-tubal insufflation. CV Mosby, St Louis Missouri

Rumery RE, Eddy EM (1974) Scanning electron microscopy of the fimbriae and ampullae of rabbit oviducts. Anat Rec 178:83–102

Seckinger DL (1923) Spontaneous contractions of the Fallopian tube of the domestic pig with reference to the oestrous cycle. Bull Johns Hopkins Hosp 34:236–239

Seckinger DL, Snyder FF (1926) Cyclic changes in the spontaneous contractions of the human Fallopian tube. Bull Johns Hopkins Hosp 39:371–378

Sjöberg N-O (1967) The adrenergic transmitter of the female reproductive tract: distribution and functional changes. Acta Physiol Scand Suppl 305:1–32

Snyder FF (1923) Changes in the Fallopian tube during the ovulation cycle and early pregnancy. Bull Johns Hopkins Hosp 34:121–125

Snyder FF (1924) Changes in the human oviduct during the menstrual cycle and pregnancy. Bull Johns Hopkins Hosp 35:141–146

Sobotta J (1895) Die Befruchtung und Furchung des Eies der Maus. Arch Mikr Anat 45:15–93

Spilman CH, Harper MJK (1973) Effect of prostaglandins on oviduct motility in estrous rabbits. Biol Reprod 9:36–45

Stalheim OHV, Gallagher JE, Deyoe BL (1975) Scanning electron microscopy of the bovine, equine, porcine and caprine uterine tube (oviduct). Am J Vet Res 36:1069–1075

Temkin O (1956) Soranus' gynecology. Hopkins, Baltimore

Thibault C (1972) Physiology and physiopathology of the Fallopian tube. Int J Fertil 17:1–13

Thiry L (1862) Über das Vorkommen eines Flimmerepithelium auf dem Bauchfell des weiblichen Frosches. Göttinger Nachrichten 171

Van Beneden E (1875) La maturation de l'oeuf, la fécondation et les premières phases du développement embryonnaire des mammifères d'après des recherches faites chez le lapin. Bull Acad Belg Cl Sci 40:686–689

Van Horne J (1668) Suarum circa partes generationis in utroque sexu observationum prodromus. Leyden

Vesalius A (1543) De humani corporis fabrica. Basel

Westman A (1926) A contribution to the question of the transit of the ovum from the ovary to the uterus in rabbits. Acta Obstet Gynecol Scand Suppl 5:1–104

Winston RML (1980) Microsurgery of the Fallopian tube: from fantasy to reality. Fertil Steril 34:521–530

Wintenberger-Torres S (1961) Mouvements des trompes et progression des oeufs chez la brebis. Ann Biol Anim Biochim Biophys 1:121–133

Wislocki GB, Guttmacher AF (1924) Spontaneous peristalsis of the excised whole uterus and Fallopian tubes of the sow with reference to the ovulation cycle. Bull Johns Hopkins Hosp 35:246–252

Woodruff JD, Pauerstein CJ (1969) The Fallopian tube: structure, function, pathology and management. Williams and Wilkins, Baltimore

CHAPTER II

Development of the Fallopian Tubes and Their Functional Anatomy

CONTENTS

Introduction 12
Embryonic Derivation of the Fallopian Tube 14
Vascular Anatomy of the Tube 18
The Lymphatic System 19
Musculature and Innervation 20
Mucosa: Cilia and Secretory Cells 24
References 27

Introduction

In many mammalian species, the regions of the Fallopian tube designated fimbriated infundibulum, ampulla and isthmus (Fig. II.1) present characteristic histological features that underlie differing physiological functions, but in the transitional regions of the ampullary-isthmic and utero-tubal junctions, the situation may be more complex. For example, in species such as pigs and rabbits, the ampullary-isthmic junction is easily recognized, but a specific junctional region is seldom seen clearly in primates (Fig. II.2), although in women it may be distinguished by palpation (Pauerstein and Eddy 1979). A further example is that the caudal extremity of the tube in primates is represented by a substantial intramural portion, whereas in a number of laboratory species, the tube enters the uterus through its mesenteric wall (Fig. II.2), and its opening is protected by a sphincter, villi or mucosal folds. Variation in the anatomy and physiology of the utero-tubal junction has been reviewed by Hafez (1973a) and Nilsson and Reinius (1969), and treated more extensively in a number of classical papers (Kelly 1927; Andersen 1928; Lee 1928). Another morphological distinction is the presence of an ovarian bursa in species such as rats and mice, in which the infundibulum is developed as a sac that more or less completely encloses the gonad (Alden 1942; Wimsatt and Waldo 1945). In other species such as cows, sheep, pigs and rabbits, the proximal end of the tube opens to the peritoneal cavity. Detailed reviews of the morphology of the Fallopian tube are available (see Hafez and Blandau 1969; Hafez 1973b).

The dimensions of the tube vary enormously between species: rather than presenting a list, examples will be taken from primates and domestic farm animals. In women, the extra-uterine portion of the tube measures approximately

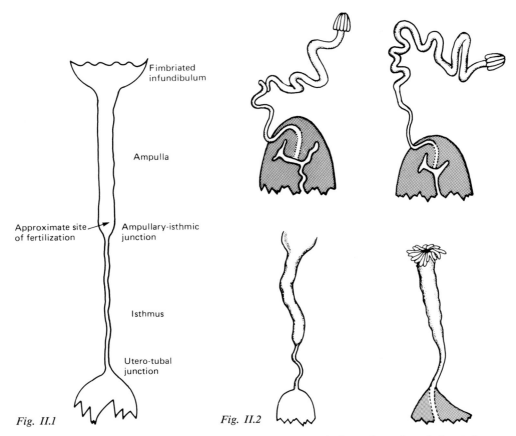

Fimbriated
infundibulum

Ampulla

Approximate site
of fertilization

Ampullary-isthmic
junction

Isthmus

Utero-tubal
junction

Fig. II.1

Fig. II.2

Fig. II.1. Diagrammatic representation of the Fallopian tube to indicate the regions designated infundibulum, ampulla and isthmus, and the relative positions of the ampullary-isthmic and utero-tubal junctions

Fig. II.2. The differing morphology of the Fallopian tubes in two rodents, an ungulate and man. (Adapted in part from Nilsson and Reinius 1969)

11 cm with the range commonly extending from 8–15 cm (Pauerstein and Eddy 1979); the intra-mural portion is 1.5–2.0 cm in length, or somewhat less if distinguished histologically (Lisa, Gioia and Rubin 1954). In horses, the tubes may commonly be 20–30 cm in length whilst in pigs, figures of 26–28 cm have been recorded, with the isthmus comprising approximately one third of the overall length of the tube (Hunter 1984). The Fallopian tubes of cows and sheep generally measure 20–25 cm and 14–16 cm, respectively, but there is considerable variation between animals for all these figures; this is influenced, not least, by the age of the animal and the stage of the ovarian cycle (Anopolsky 1928). The dimensions of the tract may continue to increase long after the age of puberty, although they are usually concealed by the convolutions of the duct in situ within its supporting mesentery, the mesosalpinx.

In terms of the microscopic structure of the Fallopian tube, it is the possibility of interactions between the gametes or embryos and tubal epithelium that presents the most exciting prospects, so rather more attention will be devoted to these aspects in the pages that follow; in particular, the distribution of cilia and secretory cells will be highlighted. However, brief reference first needs to be made to the embryonic origin of the tubes, their vascular and lymphatic systems, and then to the musculature and its innervation. These have been most recently reviewed by Rousseau, Levasseur and Thibault (1987).

Embryonic Derivation of the Fallopian Tube

The primitive embryonic structures from which the Fallopian tubes develop are illustrated in Fig. II.3. As is widely known, a female genital tract arises from the Müllerian duct system due principally to elaboration of the paramesonephric ducts. A male tract, by contrast, arises from the Wolffian duct system, with development of the mesonephric ducts. In early prenatal life, there is the potential to develop either a male or a female reproductive tract, since the very young embryo possesses both sets of primitive genital tracts.

In normal circumstances, only one set of ducts develops and the other becomes vestigial, although this may not be so in the case of chromosome anomalies or endocrine disturbance (Fig. II.4). The presence or absence of a Y-chromosome in mammals dictates which of the duct systems will develop. Genetic maleness promotes stabilization and development of the Wolffian ducts due to the influence of androgens from the embryonic testes. In this situation, regression of the Müllerian ducts is due to the influence of a Sertoli-cell protein secretion, the anti-Müllerian hormone (Josso et al. 1979; Vigier et al. 1984). In the absence of embryonic testes and male sex hormone, the Müllerian ducts are permitted to develop and the Wolffian ducts become vestigial. It is therefore primarily the absence of androgens and anti-Müllerian hormone that underlies development of the female ducts. These morphogenetic changes proceed according to an accurately-defined schedule of embryonic development (Burns 1961).

It is perhaps worth stressing at this point that, despite much detailed information, the precise manner whereby the presence of a Y-chromosome induces formation of a testis is still not clearly understood. The role of a putative H-Y antigen has been much debated (Wachtel 1983; McLaren et al. 1984), but has not found universal acceptance. Moreover, our own observations on XX intersex animals in which there is unilateral development of an ovotestis or a testis-like structure preclude systemic effects of anomalous titres of an H-Y antigen or, indeed, any simple involvement of a translocated portion of the Y-chromosome (Hunter, Cook and Baker 1985).

Elaboration of a functional Fallopian tube from the proximal portion of the Müllerian duct is, of course, a complex dynamic process. Details of the formation of individual tissues and cell types can be sought in the reviews of Price, Zaaijer and Ortiz (1969), Woodruff and Pauerstein (1969) and Hafez (1973b).

14

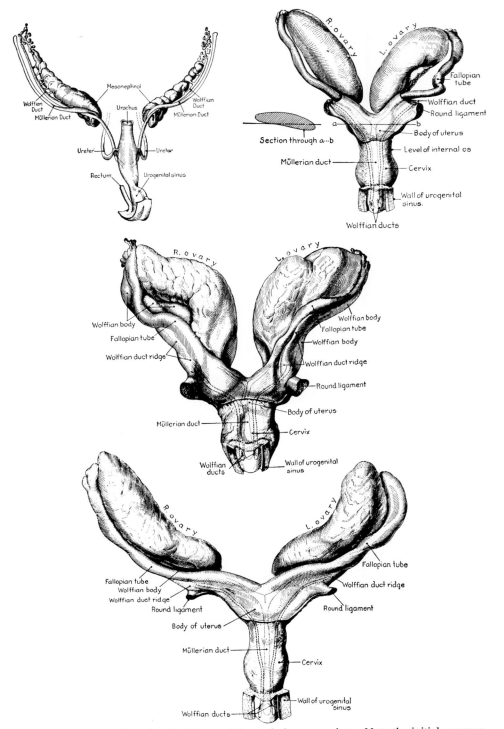

Fig. II.3. The gonads and genital ducts of the early human embryo. Note the initial presence of both Wolffian and Müllerian duct systems; and then elaboration of the Fallopian tubes from the proximal portion of the Müllerian ducts. (Adapted from Hunter 1930)

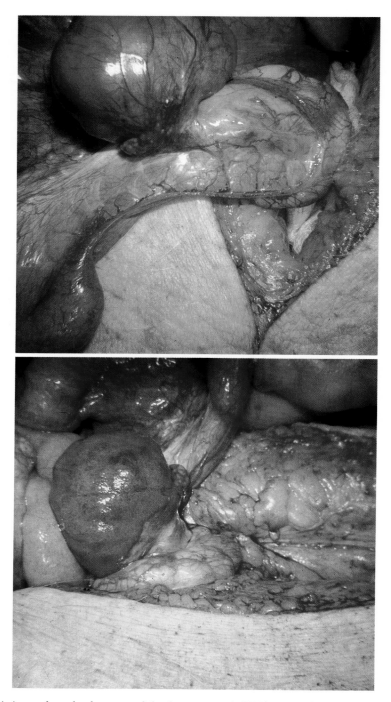

Fig. II.4. Anomalous development of the duct systems in XX intersex pigs. The gonad has differentiated as an ovotestis, with a relatively small portion of ovarian tissue, and the Fallopian tube is poorly developed, especially in the ampullary region. The Wolffian duct, by contrast, has proliferated into a prominent epididymis-like structure. (Adapted from Hunter et al. 1982, 1985)

In a majority of species, the vascular support of the Fallopian tubes is derived from both the uterine and ovarian arteries, which together furnish distinct arcades of vessels along the length of the tube (Fig. II.5). The relative contribution of the individual arterial supplies depends on the physiological situation, especially the stage of the oestrous or menstrual cycle, or the different stages of pregnancy. As judged from our own observations on the reproductive tracts of pigs, sheep and cows made during mid-ventral laparotomy, the capillary bed and larger blood vessels of the fimbria and ampulla become highly engorged close to the time of ovulation (Fig. II.6). This condition wanes only as the corpora lutea become established. The striking cyclical increase in the prominence of the vasculature appears to be promoted largely by ovarian oestrogens, and to be associated in part with an enhanced secretory or transudative function of the tube (see Chap. III). Specific correlations between the rate of blood flow to the tube and the accumulation of fluid in its lumen have yet to be presented.

A brief summary of variation between species in the arterial supply has been given by Edwards (1980), and details of the blood vessels in farm animals have been beautifully illustrated using latex-cast preparations of the tract and mesenteries for sheep, pigs and horses (Del Campo and Ginther 1973). In the guinea-pig, the ovarian artery supplies the Fallopian tube and upper portions of the uterus. By contrast, the ovarian artery is relatively small in rhesus monkeys, most of the blood supply to the reproductive tract and gonad stemming from the uterine artery, except apparently in late pregnancy (Wehrenberg et al. 1977). In women, the ovarian and uterine arteries may yield either two or three arterial branches to the tube (Woodruff and Pauerstein 1969; Pauerstein and Eddy 1979). Specific details of the vascular anatomy are given in these last reports, together with a diagrammatic representation of the arterial supply. As would be expected, the intra-mural portion of the tube is dependent upon branches from the uterine artery.

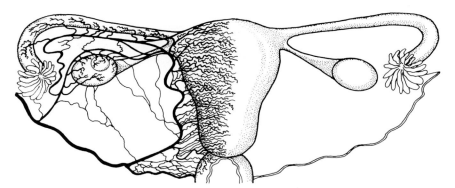

Fig. II.5. The vasculature of the human Fallopian tube to illustrate the contribution of ovarian and uterine arteries, and the disposition of arterioles from the bordering arcade. [Reconstructed from several sources, but principally from Koritké, Gillet and Piétri (1967) and Pauerstein (1974)]

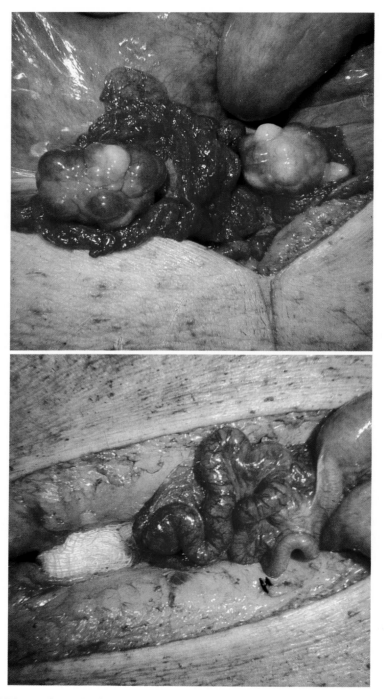

Fig. II.6. The ovaries and Fallopian tube of a domestic pig exposed at mid-ventral laparotomy to reveal pre-ovulatory Graafian follicles, some 9–10 mm in diameter, and the highly engorged folds of the fimbriated infundibulum. Note also the prominent vasculature of the convoluted ampulla in the lower part of the plate. (Adapted from Hunter 1977)

Rather more detail on the vessels associated with the human tube has been given by Gillet et al. (1967) and by Esperança Pina (1981), who note six or seven arterioles supplying the ampulla. The arterioles from the tubal arcade run through the serous and subserous coats, the diameter and degree of coiling (i. e. spiralization) of arterioles within the mucosa increasing progressively to reach a maximum during menstruation. The calibre of the capillaries and venules undergoes similar changes, but diminishes during menstruation. In pregnancy, the arterioles become more tightly coiled, whilst the overall size of arterioles, capillaries and venules increases. This microvasculature is extensively portrayed in the studies of Gillet and Piétri (1967) and Gillet and Koritké (1970).

In general, the course of the venous return closely follows that of the arterial vessels, with interconnecting capillary networks being noted at various levels in the tubal tissues – beneath the serosa, in the muscle layers, and in the mucosa. In women, these three venous plexuses become confluent in the subserous connective tissue, where the extrinsic drainage of the tube originates (see Lippes 1975). The serosal capillary network is dense in the isthmus and becomes sparse in the ampulla (Woodruff and Pauerstein 1969; Pauerstein and Eddy 1979), but this situation would doubtless also be influenced by ovarian endocrine status.

The relationship between arteries and veins in the proximity of the tube and gonad needs to be considered in the context of local counter-current transfer systems. Such a means of transferring hormones from the utero-ovarian vein to the ovarian artery is well known (McCracken, Baird and Goding 1971; Del Campo and Ginther 1973; Goding 1974; Einer-Jensen 1976), and especially concerns prostaglandin $F_{2\alpha}$ ($PGF_{2\alpha}$) produced in the uterine wall. But a similar means of shunting information from the vasculature of the Fallopian tube to that of the ovary is also worth examining. In terms of prompting ovarian responses to the presence of an embryo in the lumen of the tube, and especially in the light of accumulating evidence for early pregnancy factors (see Chap. VI), a transfer from venous blood draining the tube to the ovarian artery might act in concert with a lymphatic route.

The Lymphatic System

Quite apart from its obvious rôle in transporting fluid, the lymphatic system has attracted interest because of its potential as a means of transmitting endocrine information between neighbouring tissues or organs. Detailed studies of the lymphatic vessels in domestic animals were presented some 60 years ago (Andersen 1927), but the more recent interest (e. g. Staples, Fleet and Heap 1982) has been associated in part with examining ad-ovarian passage of uterine luteolytic factors such as $PGF_{2\alpha}$, for which the vascular counter-current transfer is quantitatively inefficient (as low as $1-2\%$; McCracken 1971; McCracken et al. 1971; McCracken 1984).

Concerning the Fallopian tubes, a network of lymphatic vessels drains the mucosa, muscularis and serosa (Fig. II.7), and then combines to course in the mesosalpinx and mesometrium to the para-aortic or lumbar nodes; thus the lymphatic vessels of the tube follow the ovarian and uterine lymphatic drainage.

19

Andersen (1927) was perhaps the first to suggest a cyclical variation in the activity of the lymphatic system, with the vessels adjoining the Fallopian tube appearing largest in the follicular phase, whilst those of the utero-tubal junction and proximal part of the uterine horn were most developed during the luteal phase (see also Morris and Sass 1966). But apart from any cyclical distinctions, Fig. II.7, representing the lymphatic system of the pig Fallopian tube, shows a more extensive network of vessels bordering the isthmus than ampulla. Since by far the greater part of the tubal residence of the embryo is within the isthmus, this lymphatic architecture could well be involved in a messenger system for transmission of embryonic information to the ovary (see above). Even so, the study of Staples et al. (1982) in sheep failed to produce evidence for a retrograde lymph flow between the uterus and ovaries, and the detailed report of Morris and Sass (1966) noted that lymphatics draining the sheep ovary were joined by lymphatics from the proximal half of the Fallopian tube. These observations do not, however, preclude counter-current transfer mechanisms for embryonic programming of ovarian events, possibly involving the ovarian artery.

A concise discussion of the lymphatic vessels associated with the human Fallopian tube is presented by Woodruff and Pauerstein (1969), and there is brief comment also in the books of Pauerstein (1974) and Edwards (1980). In the more detailed monograph of Reiffenstuhl (1964), there is specific reference to two or three lymphatic vessels emerging from the tubal wall, combining, and then coursing along the upper edge of the broad ligament to reach the lymphatic vessel network below the ovarian hilus. On the basis of his own dissections, Reiffenstuhl notes a very delicate subserous net, visible along the entire length of the tube, as well as the vessels emerging from the myosalpinx. The subserous vessels of the isthmus are connected with those of the uterine fundus and, within the subovarian plexus, the tubal lymph vessels anastomose with those of the ovary.

Musculature and Innervation

The general macroscopic nature of the musculature of the tube is depicted in transverse section in Fig. II.7, taken from Andersen's (1927) study of the domestic pig. Internal to the well-vascularized serosa are an outer layer of longitudinal muscle and an inner layer of circular muscle; for important functional reasons, the isthmus invariably has the more powerful potential, usually associated with the well-developed layer of circular muscle. Thus, the myosalpinx increases in thickness from the infundibulum to the utero-tubal junction. Arrangement of the muscular layers varies according to species, for the guinea-pig has an outer circular and an inner longitudinal layer, whilst a circular layer of muscle is disposed between outer and inner longitudinal layers of the human tube (Nilsson and Reinius 1969), at least in the distal portion forming the intra-mural segment; however, the inner longitudinal muscle disappears as a defined layer a few centimetres along the isthmus. These structural variations suggest functional differences between species in the nature of tubal contractions. In women, moreover, the musculature of the inner layer of the uterine wall at the utero-tubal junction is formed from four bundles with interlacing spiral fibres (Fig. II.8), enabling part

20

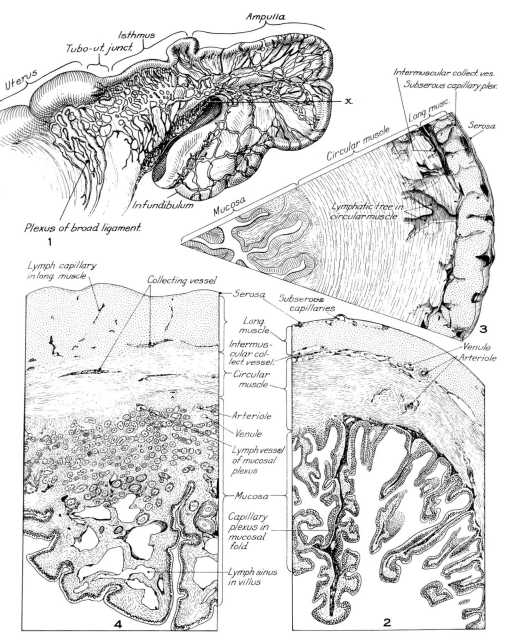

Fig. II.7. The gross morphology of the Fallopian tube of the domestic pig, with transverse sections through the ampulla (2), the isthmus (3) and the utero-tubal junction (4). The distribution of the muscular layers and lymph vessels of the mucosa is especially well illustrated (Andersen 1927)

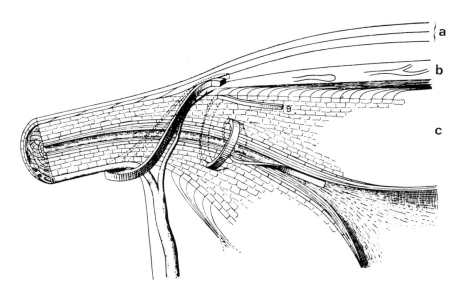

Fig. II.8. The intra-mural region of the human Fallopian tube to illustrate the complex arrangement of the smooth muscle layers that permits the formation of a physiological sphincter. (Adapted from Vasen 1959)

of the intra-mural portion of the tube to be strongly constricted (Vasen 1959; Rocker 1964). This specialization may be involved in regulating sperm transport and storage in this region of the tract (see Chaps. IV and V), and also in preventing reflux passage of menstrual or post-partum fluids.

Histochemical studies of the neuro-anatomy of the tube show innervation by a network of sympathetic fibres stemming primarily from the ovarian and/or hypogastric plexus. Although adrenergic fibres can be identified in longitudinal muscle, a significant feature of this nerve supply is that the circular muscle layers of the isthmus are very richly innervated, with high densities of adrenergic nerve terminals demonstrable in this muscular tissue by appropriate fluorescence techniques (Brundin 1965; Owman and Sjöberg 1966; Owman, Rosengren and Sjöberg 1967). At the ampullary-isthmic junction, there is a marked increase in the frequency of these nerve endings. This stands in contrast to the innervation of the ampulla and infundibulum, which is poor, and where adrenergic endings are usually restricted to the walls of the blood vessels.

Chemical studies have confirmed the fact that there is a significantly higher content of noradrenaline in the isthmus than in the ampulla (Brundin 1964, 1969; Sjöberg 1967; Hølst, Cox and Braden 1970). The dense adrenergic innervation permits the isthmus to act as a physiological sphincter which may be important for regulating the transport of gametes and embryos (see Chaps. IV, V and VII). It should be emphasized, however, that there is no histological evidence for a true anatomical sphincter in this region of the tube (Lisa et al. 1954).

In rabbits, the hypogastric nerve bundles convey to the tube all the fibres originating from the sympathetic division of the autonomic nervous system. Turning to innervation of the human Fallopian tube, this has been concisely reviewed

22

by Brundin (1969), Pauerstein (1974) and Pauerstein and Eddy (1979), with notes on both the sympathetic and parasympathetic systems (Fig. II.9). Post-ganglionic fibres pass from the inferior mesenteric ganglion to the tube via the hypogastric plexus. The isthmus and part of the ampulla are innervated by adrenergic fibres from the hypogastric plexus, whilst the proximal ampulla and fimbria receive fibres from the ovarian plexus. Most fibres of the hypogastric trunk are non-myelinated, indicating that they are post-ganglionic (Brundin 1969), but there is also experimental animal evidence for some pre-ganglionic fibres within the hypogastric nerve (Pauerstein 1974). Post-ganglionic fibres to the upper portions of the ampulla and fimbria stem from ganglia such as the aortic and renal.

Afferent tubal nerves carry pain sensation, and sensory nerves from the fimbria and ampulla may travel to the spinal cord via the ovarian plexus and splanchnic nerve, accompanying the sympathetic nerves.

The parasympathetic supply is also split, in the sense that the proximal portion of the tube is supplied by vagal fibres from the ovarian plexus, whereas the intra-mural portion receives post-ganglionic fibres from the pelvic plexus. But there is some debate as to the extent of cholinergic innervation of the myosalpinx, the consensus being that it is minimal, although the mucosa may receive cholinergic fibres. The debate has been eloquently summarized by Rousseau et al. (1987), who also offer a perspective on the rôle of vasoactive intestinal polypeptide nerves – peptidergic nerves – in the human reproductive tract.

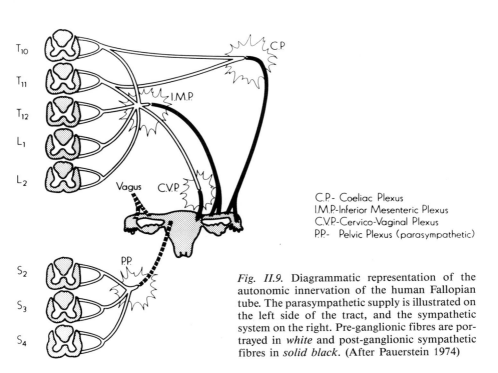

C.P.- Coeliac Plexus
I.M.P.-Inferior Mesenteric Plexus
C.V.P.-Cervico-Vaginal Plexus
P.P.- Pelvic Plexus (parasympathetic)

Fig. II.9. Diagrammatic representation of the autonomic innervation of the human Fallopian tube. The parasympathetic supply is illustrated on the left side of the tract, and the sympathetic system on the right. Pre-ganglionic fibres are portrayed in *white* and post-ganglionic sympathetic fibres in *solid black*. (After Pauerstein 1974)

23

The contours of the lumen of the Fallopian tube reflect that longitudinal folds of the mucosa, which are most accentuated in the ampulla. In fact, the arrangement of the mucosal folds generally renders the lumen only a potential space in this region; it is not as though the duct is actually open in a three-dimensional sense. The mucosal layers are of functional interest because of an extreme sensitivity to the circulating levels of ovarian steroid hormones. In the late follicular phase of the oestrous or menstrual cycle, mucosal tissues become greatly distended due to an indirect influence of oestrogens upon the prominent lymphatics (Fig. II.10), such oedema acting significantly to reduce the potential size of the tubal lumen. This is particularly so in the region of the lower isthmus and utero-tubal junction shortly before ovulation (Andersen 1927, 1928; Lee 1928), and notably in species with uterine deposition of semen such as pigs and horses; there are implications for the process of sperm transport and also for restricting the passage of uterine fluids. Although the mucosa may appear less complex in primates, it is conspicuously arranged into four primary folds at the cornual end of the tube in women (Pauerstein and Eddy 1979).

Oedema in the more intricate mucosal folds of the ampulla may also be involved in regulating the transport of gametes, fostering a suitable epithelial interaction with the egg(s) in this larger portion of the Fallopian tube. Cyclical alteration in the appearance and extent of the oedema can be mimicked by the injection of appropriate steroid hormones.

Turning to the ciliated cells of the tube, these are most prominent in the epithelial surface of the fimbriated infundibulum where they form dense arrays. Large numbers of cilia may also be found in the ampulla, although non-ciliated secretory cells are equally abundantly distributed throughout this portion of the tube in many species. In general, ciliated cells are less frequent in the isthmus, but they have been reported in significant numbers in sheep (Hadek 1955), rabbit and man (Nilsson and Reinius 1969; Patek 1974), pig (Gaddum-Rosse and Blandau 1973; Wu, Carlson and First 1976) and cow (Gaddum-Rosse and Blandau 1976); ciliated cells are extremely scarce, however, in the isthmus of guinea-pigs and rats (Gaddum-Rosse and Blandau 1976). They may be present on the surface of the utero-tubal junction (Wu et al. 1976; Fléchon and Hunter 1981; Hunter, Fléchon and Fléchon 1987), and in the intra-mural portion of the human tube (Fadel et al. 1976).

Growth of the ciliated epithelium appears to be under the influence of ovarian hormones (Allen 1928), and data from the rabbit, monkey and man all indicate that growth of these ciliated cells is promoted by oestrogens. Treatment with progesterone, by contrast, antagonizes the oestrogen-driven cell growth (Brenner 1969a). In at least one species of primate, the rhesus monkey, the infundibular cilia are dependent on oestrogen, disappearing during the luteal phase and undergoing regeneration between succeeding menstrual cycles (Brenner 1969b). This differs from the situation in the ampullary epithelium in which there is no dramatic loss of cilia; the principal difference is thought to be one of sensitivity to gonadal hormones between these two regions of the tube. No significant degeneration and regeneration of cilial structures has been detected during the

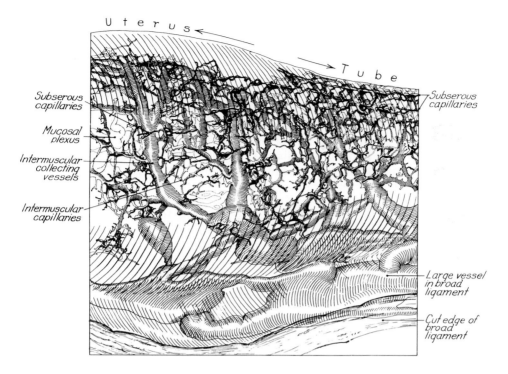

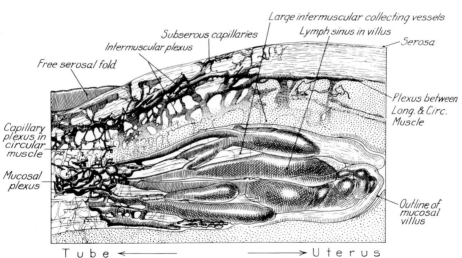

Fig. II.10. A detailed view of the region of the utero-tubal junction in a domestic pig to emphasize the prominent lymphatic vessels that serve to dilate the polypoid processes during oestrus. This arrangement prevents the free passage of fluids – such as semen – from the lumen of the uterus into that of the Fallopian tube. (After Andersen 1927)

menstrual cycle of women (Hashimoto et al. 1964; Patek 1974), although deciliation of the tubal isthmus is well known during the last stages of pregnancy.

The rate of beat of the cilia is largely dictated by the circulating level of ovarian hormones, activity being greatest at ovulation or shortly afterwards when the stroke of the cilia in the fimbriated portion of the tube is closely synchronized and direction-orientated towards the ostium (Borell, Nilsson and Westman 1957; Blandau 1969). This enables the egg(s) within an investment of cumulus cells to be stripped from the apex of collapsing follicles and rapidly propelled into the ampulla. In all mammals so far examined, including the rat, rabbit, cat, cow, sheep, monkey and human, the cilia in the ampulla beat in the direction of the ampullary-isthmic junction at this time (Blandau and Verdugo 1976), suggesting a functional rôle in the process of egg transport to the site of fertilization. However, any overriding contribution of cilial activity to egg transport in the tubes has been questioned since, in women suffering from the immotile cilia syndrome, pregnancies have occasionally been reported (see Chaps. IV and X).

Secretory cells are well distributed throughout the ampulla of the Fallopian tubes in higher mammals. However, there seems to be some controversy as to the extent to which such cells are also found in the isthmus (see Whittingham 1968; Nilsson and Reinius 1969). Studies with the scanning electron microscope indicate that secretory cells are present, at least in the isthmus of domestic ungulates (Wu et al. 1976; Fléchon and Hunter 1981; Hunter et al. 1987) and primates (Patek 1974), especially on the ridges of the longitudinal folds. It is within the isthmus, moreover, that the developing rabbit embryo becomes invested with a substantial layer of mucin which clearly originates from the tubal epithelium (see Greenwald 1958), and the presence of tenacious mucus in the human isthmus has received considerable emphasis (Patek 1974; Jansen 1980). In fact, Pauerstein and Eddy (1979) have remarked that secretory function is more prominent in the human isthmus than ampulla.

The cytology of the secretory cells changes throughout the oestrous or menstrual cycle, epithelial height and secretory activity reaching a maximum close to the time of ovulation (see Reinius 1970; Patek 1974; Chap. III). During the follicular phase in women, these cells have well-developed and evenly-distributed microvilli (Patek 1974). Secretory material accumulated in epithelial cells during the follicular phase of the cycle is evacuated into the lumen after ovulation, and epithelial height then decreases. These variations generally become less pronounced moving from the infundibulum to the isthmus (Hafez 1973 a). Secretory granules are found within various non-ciliated-cells, and their occurrence may differ in the same cell type during different reproductive states, since the specific liberation of granules is also programmed by the gonadal hormones. A question of increasing relevance is the extent to which the developing embryo can influence the secretory activity of the tubal epithelium. The physiological situation may be a gonadal regulation of overall epithelial activity in conjunction with a local 'fine-tuning' by embryonic signals.

REFERENCES

Alden RH (1942) The periovarial sac in the albino rat. Anat Rec 83:421–435

Allen E (1928) Further experiments with an ovarian hormone in the ovariectomized adult monkey, *Macacus rhesus*, especially the degenerative phase of the experimental menstrual cycle. Am J Anat 42:467–487

Alm P, Alumets J, Hakanson R, Helm G, Owman C, Sjoberg N-O, Sundler F (1980) Vasoactive intestinal polypeptide nerves in the human female genital tract. Am J Obstet Gynecol 136:349–351

Andersen DH (1927) Lymphatics of the Fallopian tube of the sow. Contrib Embryol Carneg Instn 19:135–148

Andersen DH (1928) Comparative anatomy of the tubo-uterine junction. Histology and physiology in the sow. Am J Anat 42:255–305

Anopolsky D (1928) Cyclic changes in the size of muscle fibers of the Fallopian tube of the sow. Am J Anat 40:459–469

Blandau RJ (1969) Gamete transport – comparative aspects. In: Hafez ESE, Blandau RJ (eds) The mammalian oviduct. University of Chicago Press, Chicago, pp 129–162

Blandau RJ, Verdugo P (1976) An overview of gamete transport-comparative effects. In: Harper MJK et al. (eds) Symposium on ovum transport and fertility regulation. Scriptor, Copenhagen, pp 138–146

Borell U, Nilsson O, Westman A (1957) Ciliary activity in the rabbit Fallopian tube during oestrus and after copulation. Acta Obstet Gynecol Scand 36:22–28

Brenner RM (1969a) The biology of oviductal cilia. In: Hafez ESE, Blandau RJ (eds) The mammalian oviduct. University of Chicago Press, Chicago, pp 203–229

Brenner RM (1969b) Renewal of oviduct cilia during the menstrual cycle in the rhesus monkey. Fertil Steril 20:599–611

Brundin J (1964) The distribution of noradrenaline and adrenaline in the Fallopian tube of the rabbit. Acta Physiol Scand 62:156–159

Brundin J (1965) Distribution and function of adrenergic nerves in the rabbit Fallopian tube. Acta Physiol Scand 66, Suppl 259:1–57

Brundin J (1969) Pharmacology of the oviduct. In: Hafez ESE, Blandau RJ (eds) The mammalian oviduct. University of Chicago Press, Chicago, pp 251–269

Burns RK (1961) Role of hormones in the differentiation of sex. In: Young WC (ed) Sex and internal secretions, 3rd edn. Williams & Wilkins, Baltimore, vol 1, pp 76–158

Del Campo CH, Ginther OJ (1973) Vascular anatomy of the uterus and ovaries and the unilateral luteolytic effect of the uterus: horses, sheep and swine. Am J Vet Res 34:305–316

Edwards RG (1980) Conception in the human female. Academic Press, London

Einer-Jensen N (1976) Vascular counter current exchange in the male and female reproductive systems. Karger Gazette 33:5–7

Esperança Pina JAR (1981) Microvascularization of the human ovary and uterine tube. In: Semm K, Mettler L (eds) Proc 3rd World Congr Human Reproduction. Excerpta Medica, Amsterdam, pp 36–46

Fadel H, Berns D, Zaneveld LJD, Wilbanks GD, Brueschke EE (1976) The human utero-tubal junction: a scanning electron microscope study during different phases of the menstrual cycle. Fertil Steril 27:1176–1186

Fléchon JE, Hunter RHF (1981) Distribution of spermatozoa in the utero-tubal junction and isthmus of pigs, and their relationship with the luminal epithelium after mating: a scanning electron microscope study. Tissue & Cell 13:127–139

Gaddum-Rosse P, Blandau RJ (1973) In vitro studies on ciliary activity within the oviducts of the rabbit and pig. Am J Anat 136:91–104

Gaddum-Rosse P, Blandau RJ (1976) Comparative observations on ciliary currents in mammalian oviducts. Biol Reprod 14:605–609

Gillet JY, Koritké JG (1970) La microvascularisation de la trompe utérine chez la femme. Arch Anat 53:125–142

Gillet JY, Piétri J (1967) Contribution à l'étude de la vascularisation de la trompe de Fallope. Aspects microscopiques. Rev Fr Gynécol 62:129–141

Gillet JY, Leissner P, Koritké JG, Muller P (1967) Considérations sur les variations de la microvascularisation au niveau de la muqueuse tubaire chez la femme pendant le cycle ovarien, la gravidité et la menopause. CR Assn Anat 52:571−581

Goding JR (1974) The demonstration that $PGF_{2\alpha}$ is the uterine luteolysin in the ewe. J Reprod Fertil 38:261−271

Greenwald GS (1958) Endocrine regulation of the secretion of mucin in the tubal epithelium of the rabbit. Anat Rec 130:477−495

Gruber CM (1933) The autonomic innervation of the genito-urinary system. Physiol Rev 13:497−609

Hadek R (1955) The secretory process in the sheep's oviduct. Anat Rec 121:187−205

Hafez ESE (1973a) Anatomy and physiology of the mammalian uterotubal junction. In: Greep RO, Astwood EB (eds) Handbook of physiology. Endocrinology II, Part II. American Physiological Society, Washington DC, pp 87−95

Hafez ESE (1973b) Endocrine control of the structure and function of the mammalian oviduct. In: Greep RO, Astwood EB (eds) Handbook of physiology. Endocrinology II, Part II. American Physiological Society, Washington DC, pp 97−122

Hafez ESE, Blandau RJ (eds) (1969) The mammalian oviduct. University of Chicago Press, Chicago

Hashimoto M, Shimoyama T, Kosaka M, Komori A, Hirasawa T, Yokoyama Y, Akashi K (1964) Electron microscopic studies on the epithelial cells of the human Fallopian tube. J Jpn Obstet Gynaecol Soc 11:92−100

Hølst PJ, Cox RI, Braden AWH (1970) The distribution of noradrenaline in the sheep oviduct. Aust J Exp Biol Med Sci 48:563−565

Hunter RH (1930) Observations on the development of the human female genital tract. Contrib Embryol 22 No 129:91−107

Hunter RHF (1977) Function and malfunction of the Fallopian tubes in relation to gametes, embryos and hormones. Europ J Obstet Gynecol Reprod Biol 7:267−283

Hunter RHF (1984) Pre-ovulatory arrest and peri-ovulatory redistribution of competent spermatozoa in the isthmus of the pig oviduct. J Reprod Fertil 72:203−211

Hunter RHF, Baker TG, Cook B (1982) Morphology, histology and steroid hormones of the gonads in intersex pigs. J Reprod Fertil 64:217−222

Hunter RHF, Cook B, Baker TG (1985) Intersexuality in five pigs, with particular reference to oestrous cycles, the ovotestis, steroid hormone secretion and potential fertility. J Endocrinol 106:233−242

Hunter RHF, Fléchon B, Fléchon JE (1987) Pre- and peri-ovulatory distribution of viable spermatozoa in the pig oviduct: a scanning electron microscope study. Tissue & Cell 19:423−436

Jansen RPS (1980) Cyclic changes in the human Fallopian tube isthmus and their functional importance. Amer J Obstet Gynecol 136:292−308

Josso N, Picard JY, Dacheux JL, Courot M (1979) Detection of anti-Müllerian activity in boar rete testis fluid. J Reprod Fertil 57:397−400

Kelly GL (1927) The uterotubal junction in the guinea-pig. Am J Anat 40:373−385

Koritké JG, Gillet JY, Piétri J (1967) Les artères de la trompe utérine chez la femme. Arch Anat Histol Embryol Norm Exptl 40:49−70

Lee FC (1928) The tubo-uterine junction in various animals. Bull John Hopkins Hosp 42:335−357

Lippes J (1975) Applied physiology of the uterine tube. Obstet Gynecol Ann 4:119−166

Lisa JR, Gioia JD, Rubin IC (1954) Observations on the interstitial portion of the Fallopian tube. Surgic Gynecol Obstet 99:159−169

McCracken JA (1971) Prostaglandin $F_{2\alpha}$ and corpus luteum regression. Ann NY Acad Sci 180:456−472

McCracken JA (1984) Update on luteolysis-receptor regulation of pulsatile secretion of prostaglandin $F_{2\alpha}$ from the uterus. Res Reprod 16:1−2

McCracken JA, Baird DT, Goding JR (1971) Factors affecting the secretion of steroids from the transplanted ovary of the sheep. Recent Progr Horm Res 27:537−582

McLaren A, Simpson E, Tomonari K, Chandler P, Hogg H (1984) Male sexual differentiation in mice lacking H-Y antigen. Nature (Lond) 312:552−555

Morris B, Sass MB (1966) The formation of lymph in the ovary. Proc R Soc B 164:577−591

Nilsson O, Reinius S (1969) Light and electron microscopic structure of the oviduct. In: Hafez ESE, Blandau RJ (eds) The mammalian oviduct. University of Chicago Press, Chicago, pp 57–83

Owman CH, Sjöberg N-O (1966) Adrenergic nerves in the female genital tract of the rabbit: with remarks on cholinesterase-containing structures. Z Zellforsch 74:182–197

Owman CH, Rosengren E, Sjöberg N-O (1967) Adrenergic innervation of the human female reproductive organs: a histochemical and chemical investigation. Obstet Gynecol 30:763–773

Patek E (1974) The epithelium of the human Fallopian tube. Acta Obstet Gynecol Scand 53 Suppl 31:1–28

Pauerstein CJ (1974) The Fallopian tube: a reappraisal. Lea & Febiger, Philadelphia

Pauerstein CJ, Eddy CA (1979) Morphology of the Fallopian tube. In: Beller FK, Schumacher GFB (eds) The biology of the fluids of the female genital tract. Elsevier, North-Holland, pp 299–317

Price D, Zaaijer JJP, Ortiz E (1969) Prenatal development of the oviduct in vivo and in vitro. In: Hafez ESE, Blandau RJ (eds) The mammalian oviduct. University of Chicago Press, Chicago, pp 29–46

Reiffenstuhl G (1964) The lymphatics of the female genital organs. Lippincott, Philadelphia

Reinius S (1970) Morphology of oviduct, gametes and zygotes as a basis of oviductal function in the mouse. I Secretory activity of oviductal epithelium. Int J Fertil 15:191–209

Rocker I (1964) The anatomy of the utero-tubal junction area. Proc Roy Soc Med 57:707–709

Rousseau JP, Levasseur MC, Thibault C (1987) L'isthme et la jonction utéro-tubaire. Contraception, Fertil Sexualité 15:227–239

Sjöberg N-O (1967) The adrenergic transmitter of the female reproductive tract: distribution and functional changes. Acta Physiol Scand Suppl 305:1–32

Staples LD, Fleet IR, Heap RB (1982) Anatomy of the utero-ovarian lymphatic network and the composition of afferent lymph in relation to the establishment of pregnancy in the sheep and goat. J Reprod Fertil 64:409–420

Vasen LCLM (1959) The intramural part of the Fallopian tube. Int J Fertil 4:309–314

Vigier B, Tran D, Legeai L, Bézard J, Josso N (1984) Origin of anti-Müllerian hormone in bovine freemartin fetuses. J Reprod Fertil 70:473–479

Wachtel SS (1983) H-Y antigen and the biology of sex determination. Grune & Stratton, New York

Wehrenberg WB, Chaichareon DP, Dierschke DJ, Rankin JH, Ginther OJ (1977) Vascular dynamics of the reproductive tract in the female Rhesus monkey: relative contributions of ovarian and uterine arteries. Biol Reprod 17:148–153

Whittingham DG (1968) Development of zygotes in cultured mouse oviducts. J Exp Zool 169:391–398

Wimsatt WA, Waldo CM (1945) The normal occurrence of a peritoneal opening in the bursa ovarii of the mouse. Anat Rec 93:47–53

Woodruff JD, Pauerstein CJ (1969) The Fallopian tube. Structure, function, pathology and management. Williams & Wilkins, Baltimore

Wu ASH, Carlson SD, First NL (1976) Scanning electron microscopic study of the porcine oviduct and uterus. J Anim Sci 42:804–809

CHAPTER III

Fallopian Tube Fluid: The Physiological Medium for Fertilization and Early Embryonic Development

CONTENTS

Introduction   30
Dynamics of Tubal Fluid Accumulation   31
Direction of Fluid Flow   33
Composition of Tubal Fluids   34
Regional and Microenvironments   43
Hormones Detected in Tubal Fluids   45
References   48

Introduction

In the absence of pathological conditions, fluid can be detected in the lumen of the Fallopian tube throughout the reproductive lifespan. It varies in volume and composition according to the stage of the ovarian cycle or pregnancy, and is most abundant at the time when gametes or embryos would be present in the tubes. Accordingly, tubal fluid might be viewed as of critical importance in the events of fertilization and subsequent cleavage of the zygote, not least since there are no fixed morphological relationships between the embryo and the endosalpinx during its progression towards the uterus. On the other hand, studies of in vitro fertilization and the first stages of development using supposedly chemically-defined media − in conjunction with subsequent transplantation of the embryos to demonstrate viability − place a substantial question mark over any unique value of the Fallopian tube fluid, this being especially so in primates. The dilemma will be addressed in Chapters IX and X.

Interest has also focussed on tubal fluids since, as will be noted, they are extremely sensitive quantitatively and qualitatively to modifications in the concentration of circulating gonadal steroid hormones. Disruption of fluid secretion and composition as a sequel to steroid treatments for synchronizing the oestrous cycles of farm animals may lead to significant depression of conception rates in the reprogrammed cycle (Hunter 1980). In our own species, post-coital contraceptive therapy involving steroid hormones may frequently act through modifications of the tubal milieu rather than by means of a classical negative feedback action to inhibit ovulation. Gonadal hormones are known to influence both the extent and activity of the vascular bed (i.e. rate of blood flow) and the epithelial cell layers of the tract; changes in the latter can be demonstrated histologically.

The effectiveness, advantages and disadvantages of the various methods of fluid collection from the lumen of the Fallopian tube have been reviewed by David, Serr and Czernobilsky (1973), Johnson and Foley (1974), Bazer, Roberts and Sharp (1978) and Mastroianni and Go (1979).

Fluid in the lumen of the Fallopian tubes is formed by (1) selective transudation from the blood and (2) active secretion from the endosalpinx. It is well established, moreover, that the epithelial secretory cells undergo pronounced changes during oestrous or menstrual cycles and, as inferred above, accumulation of fluid in the lumen is programmed by the concentration of oestrogens and progesterone circulating in the blood. The volume of fluid is greatest around the period of oestrus and near to or just after the time of ovulation, and minimal during the mid-luteal phase or in pregnancy (Table III.1). Increasing accumulation of tubal fluid is closely correlated with the phase of dilatation in the neighbouring lymphatic system (see Chap. II). Secretory pressure is also apparently greatest at the time of oestrus (Bishop 1956), although this may in part be a reflection of the oedematous condition of the mucosa together with enhanced tonicity of the myosalpinx. The influence of ovarian steroid hormones on secretion is clearly demonstrated after ovariectomy of oestrous rabbits, when tubal fluid secretion decreases from 1.2 ml to 0.2 ml/24 h; systemic injections of oestradiol will restore fluid production to oestrous volumes (Bishop 1956), highlighting the predominant rôle of oestrogens in regulating fluid production.

Comparable observations have been made in sheep (Fig. III.1). Fluid accumulation in the tubal lumen is at a maximum 24 h after the onset of oestrus, reaching values of $>1.3-1.4$ ml/24 h, but it declines to 0.05 to 0.09 ml/24 h

Table III.1. Examples of the range in rate of fluid secretion into the lumen of the Fallopian tube of various species at oestrus and during the luteal phase of the cycle (figures represent ml of fluid/24 h) (Hunter 1977)

Species	Reproductive state of animal		Reference
	Oestrus (or midcycle peak)	Luteal phase (or pregnancy*)	
Rabbit	0.8	0.3*	Bishop (1956)
	1.4	–	Clewe and Mastroianni (1960)
	2.0	0.4	Hamner and Williams (1965)
Sheep	1.3	0.3	Perkins et al. (1965)
	1.0	0.4	Restall (1966)
	1.2	0.2	Bellve and McDonald (1968, 1970)
	0.9	0.2	Roberts et al. (1976)
Cow	4.1	0.7	Stanke et al. (1973)
	2.0	0.2	Roberts et al. (1975)
Rhesus monkey	2.7	0.2	Mastroianni et al. (1961)
	2.1	0.3	Yoshinaga et al. (1971)
	1.9	0.7[a]	Mastroianni et al. (1973)

[a] The latest sample in this study was taken on Day 17 of the menstrual cycle.

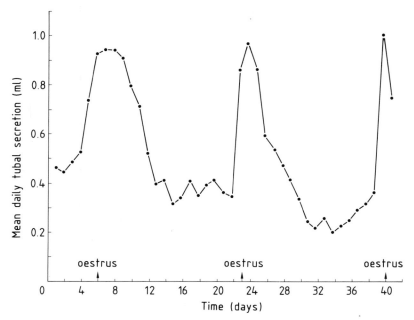

Fig. III.1. A graph of daily fluid accumulation within the lumen of the Fallopian tubes of mature ewes during two spontaneous oestrous cycles. Maximum volumes of fluid are found close to the time of ovulation (Restall 1966)

following ovariectomy; this fall can be counteracted by systemic administration of oestradiol benzoate (Restall 1966; McDonald and Bellve 1969), whereas progesterone alone does not restore fluid production in ovariectomized ewes. Seasonal influences on flow rates have been noted in sheep, with greatest volumes of fluid being collected early in the breeding season (Murray, Goode and Linnerud 1969; Sutton et. al. 1984). Figure III.2 depicts cyclical changes in the volume of tubal fluid collected from a mature cow, ranging from 0.2 ml/24 h in the luteal phase to 2 ml/24 h at oestrus (Roberts, Parker and Symonds 1975). Another set of values is presented in Table III.1, in which study a maximum volume of 6.9 ml/24 h was recorded during oestrus (Stanke et al. 1973). The earlier work of Carlson, Black and Howe (1970), using small cows, reported oestrous fluid volumes of 1.2 – 1.7 ml/24 h. In pigs, Iritani, Sato and Nishikawa (1974) have recorded accumulation rates for tubal fluid of 6.3 ml/24 h during oestrus falling to 2.1 ml/24 h in the luteal phase. The volume of fluid was highest on the second day of oestrus, which is the day of ovulation in this species.

As to primates, the best evidence is available for rhesus monkeys (Table III.1) and, again, a peak of fluid production is reached close to the time of ovulation, this of course occurring in the middle of the menstrual cycle (Yoshinaga, Mahoney and Pincus 1971). Whilst Pauerstein (1974) suggests that no data exist on rates of secretion or on specific hormonal influences on the volume and composition of human tubal fluid, Hamner (1973) in a major chapter offers the figure of 1.3 ml/24 h for fluid accumulation in women. In a shorter review, Mastroianni

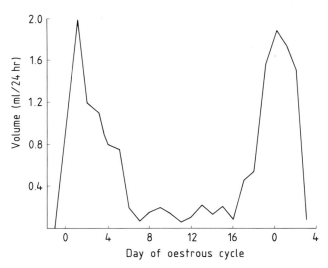

Fig. III.2. Cyclic variation in the rate of fluid accumulation within the Fallopian tube of a mature cow, indicating peak volumes around the time of ovulation (Roberts et al. 1975)

and Go (1979) noted that the rate of human tubal fluid production increases significantly during the late proliferative phase of the menstrual cycle. Although not easily justifiable from an ethical standpoint, research in this area might well have clinical value, especially in the current era of procedures of fertilization and early embryonic development in vitro, as well as in connection with cryobiological studies. In fact, an early step in this direction was the paper by Lippes et al. (1972) dealing with the collection and analysis of human tubal fluid. In subsequent papers, Lippes (1975, 1979) and Lippes et al. (1981) reported that the largest volumes of fluid were obtained around the time of ovulation, coincident with the oestrogen peak. One patient in his series produced the exceptional volume of 24 ml of fluid from the two tubes in a period of 24 h (Lippes 1979). This can scarcely be viewed as physiological, and may in part represent a response to the catheterisation technique. More conventional mid-cycle volumes were 3.9 ml/24 h (Lippes et al. 1981).

Direction of Fluid Flow

Concerning the direction of flow, the bulk of the fluid passes from the ampulla into the peritoneal cavity at oestrus in species lacking an ovarian bursa; such retrograde flow is said to account for up to 90% of the fluid in rabbits, sheep and cows (Hamner 1973). However, by the time embryos descend through the utero-tubal junction into the uterus, much of the reduced volume of fluid is being transmitted in the opposite direction into the uterine lumen (Bellve and McDonald 1970; Fig. III.3). One indication of the direction of flow can be obtained simply by ligating the proximal end of the ampulla in oestrous animals: fluid accumulates in the lumen until the stage of the cycle at which embryos would nor-

33

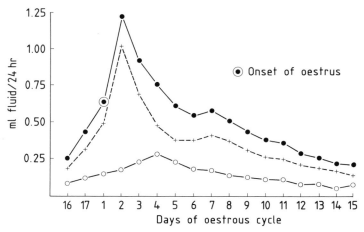

Fig. III.3. Rates of Fallopian tube fluid accumulation in sheep, and the direction of flow from the tube during the first two cycles of the breeding season. (+is via the ampulla; o is via the isthmus) (Bellve and McDonald 1970)

mally pass into the uterus, whereupon the fluid distension is rapidly terminated (Woskressensky 1891; Black and Asdell 1959; Black and Davis 1962; Hamner 1973). The directional causes of fluid flow in an undulating duct lumen may be complex, but a principal factor in the Fallopian tube is thought to be the phase of oedema in the isthmic mucosa; this is largely programmed by oestradiol and wanes as circulating levels of plasma progesterone increase with the post-ovulatory interval. Ciliary currents and the degree of muscular stricture in the isthmus and utero-tubal junction may also influence the direction of flow, and again are regulated by gonadal steroid hormones.

The nature of interactions between the volume of fluid, its direction of flow, and movement of the gametes is uncertain; the dynamics of these relationships are difficult to envisage. Similarly, a specific rôle for the tubal fluid in promoting the transport of embryos from the isthmus into the uterus remains to be demonstrated. The preceding remarks are concerned with bulk flow of fluids in the Fallopian tubes, but it is important also to consider a microflow system close to the epithelium and within the grooves or channels of the undulating surface (Fig. III.4). Microflow of tubal fluid may differ in direction from that of the bulk contents, and could be intimately involved with the movement of spermatozoa. This has been suspected at least since the writings of Parker (1931), but has yet to be convincingly demonstrated.

Composition of Tubal Fluids

Diverse aspects of tubal fluid composition and its cyclic changes have been reported. Reviews by Hamner (1973), Stone and Hamner (1975) and Menezo (1979) are suggested as background reading, although many valuable studies have since been

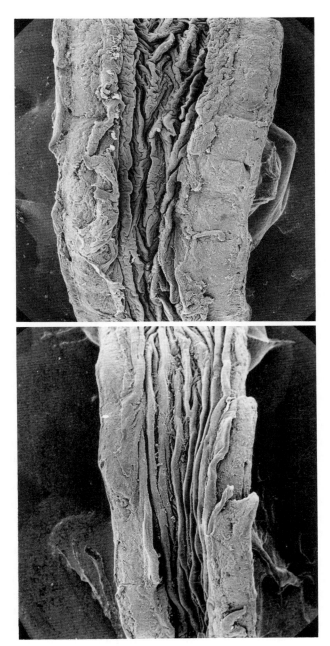

Fig. III.4. A scanning electron micrograph of portions of the Fallopian tube isthmus prepared from a mature pig near to the time of ovulation. A microflow system of fluid may be found close to the epithelial surface within the grooves or channels between the longitudinal folds of tissue. (Courtesy of Dr. J. E. Fléchon)

35

published. Apart from specific secretion from the epithelial tissues and selective transudation from the vascular bed, contributions to the tubal fluid may come from (1) the contents of Graafian follicles released at ovulation, (2) peritoneal fluid in species lacking an ovarian bursa, and (3) uterine fluid when reflux movement of liquid through the utero-tubal junction is possible. The proportion of fluid derived from each of these sources will vary according to the stage of the ovarian cycle.

Protein and Amino Acid Content. The protein content of fluid from the Fallopian tubes has been measured in numerous experiments (see Beier 1974; Aitken 1979). The somewhat variable results should not cause undue surprise, since these have frequently been based upon different procedures of sample collection. None could be viewed as strictly physiological. But, as an overall statement, the concentration of proteins in tubal fluid is usually low when compared with that in blood. This would be in line with the evidence for a selective passage of macromolecules from the bloodstream into the tubal lumen (Glass and McClure 1965; Glass 1969).

Restall and Wales (1966) reported values of 1.36 g protein/100 ml in sheep tubal fluid, compared with a mean value of 3.02 g/100 ml in the study of Perkins and Goode (1966). Even larger variations in the protein content of tubal fluid have been noted between individual cows. For example, Carlson et al. (1970) found protein concentrations of 0.08 and 0.16 g/100 ml in fluid collected from two cows, Stanke et al. (1974) reported values of 4.0 − 5.2 g/100 ml, and Roberts et al. (1975) noted a range from 0.3 − 3.2 g/100 ml. Despite this considerable range in protein concentration, there was no significant variation on the basis of the stage of the oestrous cycle − perhaps indicating that the bulk of the protein constituents of tubal fluid stem from a serum transudate. Such was the conclusion of Roberts et al. (1975) from their study in cows.

The latter point is one of major interest − the extent to which specific epithelial secretion adds to and modifies the composition of a transudate from the blood. Gel electrophoresis analysis of proteins from human tubal fluid reveals patterns that are similar to those of the corresponding serum, but some unique proteins such as β-glycoprotein are also revealed in women (Moghissi 1970), in monkeys (Mastroianni, Urzua and Stambaugh 1970), and seemingly in sheep (Roberts, Parker and Symonds 1976). More recent work on glycoproteins in sheep tubal fluid has indicated a novel substance, subunit size of M_r 80 − 90000, that is present for 3 to 6 days of each oestrous cycle, coinciding with the period of a high rate of fluid flow and the presence of embryos in the Fallopian tube (Sutton et al. 1984). The protein first appeared on the day of oestrus or the following day (i.e. the day of ovulation) and was one of the two major proteins as determined by PAS staining. In a subsequent paper on sheep fluid, the subunit of the glycoprotein was ascribed an M_r of 70 − 90000, and it was noted that the protein had the ability to bind to spermatozoa, suggesting a rôle in the events of fertilization (Sutton, Nancarrow and Wallace 1986). A follicular phase, oestrogen-dependent protein of M_r 130000 has recently been reported in the tubal fluid of baboons (Fazleabas and Verhage 1986).

The protein content of human tubal fluid has been examined in specimens collected via mini-laparotomy and catheterisation in patients undergoing voluntary sterilization (Lippes et al. 1981). Protein concentrations were lowest shortly before ovulation, when fluid volumes were greatest, and highest immediately before and after menstruation. In a large proportion of the patients, specific proteins appeared in tubal fluid at the time of ovulation and receded or disappeared within 3 to 5 days. Electrophoresis revealed these proteins to migrate in the albumin and β-globulin region. No firm evidence for the β-glycoprotein of Moghissi (1970) was offered, but Fazleabas and Verhage (1986) presented preliminary evidence for a human follicular phase protein similar to the 130000 M_r protein in baboon tubal fluid. The distribution of major proteins in human tubal fluid is similar to that in blood (Pauerstein 1974); α- and γ-globulins are included, with γ-globulin appearing as the principal immunoglobulin. Relevant quantitatively is the finding that, when rhesus monkeys were immunized with a model antigen, there was 10- to 40-fold less antibody in tubal fluid when compared with serum (Schumacher, Yang and Broer 1979).

In rabbit tubal fluid, an oestrogen-modulated protein with the electrophoretic migration properties of uteroglobin has been described (Feigelson and Kay 1972; Beier 1974), and a large glycoprotein (subunit size M_r 73000) has been reported in fluid from oestrous rabbits which was thought to be involved in the egg coating process (Shapiro, Brown and Yard 1974); this may correspond to the high molecular weight glycoprotein of Stone, Huckle and Oliphant (1980) detected in tubal fluid during pseudopregnancy or after progesterone administration to ovariectomized does. Non-serum proteins have been detected in cow tubal fluid close to the time of ovulation and during dioestrus (Roberts et al. 1975). Although results to date still suggest that most proteins in tubal fluid arise from a serum transudate, there may be a cyclical variation in the ratio between transudate and specifically-synthesized secretions (Mastroianni et al. 1970; Brackett and Mastroianni 1974). Moreover, the low concentrations of protein in tubal fluid, as compared with those in serum, could indicate that transudation accounts for only a small part of the actual fluid and that the bulk of the aqueous phase results from an active secretory process (Roberts et al. 1975).

Significant for this discussion is the observation that Fallopian tubes from oestrous rabbits can be cultured in vitro to synthesize and secrete specific sulphated glycoproteins; the type and amount of secretion differ between the ampullary and isthmic portions (Oliphant et al. 1984; Hyde and Black 1986).

Histochemical examination of the epithelium from cow Fallopian tubes has revealed the presence of PAS-positive material on the surface of non-ciliated cells and in secretory granules (Björkman and Fredricsson 1966; Fredricsson 1969). Because this staining was resistant to prior treatment with diastase, glycoproteins or glycopeptides were thought to be responsible. However, biochemical examination of tubal secretion failed to reveal the presence of appreciable amounts of epithelial glycoprotein, thereby suggesting only minor amounts of such secretions (Roberts et al. 1975). Over and above these details, it remains questionable just how important the protein content of tubal fluid is for the normal functioning of the gametes and/or embryos. Unique proteins may facilitate sperm – egg interactions. But, as far as embryos are concerned, Krebs cycle intermediaries seem to be the critical substrates for early development (see Chap. VI). Although large molecular weight substances may be essential to protect the plasma membrane of the gametes – especially of spermatozoa – and also of the embryos. Rabbit embryos are well known to become invested with a mucin layer during their passage along the Fallopian tubes, a phenomenon which is doubtless related to the post-ovulatory release into the tubal lumen of epithelial granules staining for acid mucopolysaccharides (Braden 1952). Mouse embryos have also been reported to become invested with a large glycoprotein during early development, but it is not certain whether this is a specific secretory product of the endosalpinx (Kapur and Johnson 1985).

Diverse enzymes can be detected histochemically in the epithelium of the Fallopian tube (Fredricsson 1969), but their presence and activity in the luminal fluids are by no means fully defined (Hamner 1973). Again, however, one might expect specific products of reproductive tract tissues to be regulated by gonadal steroid hormones and therefore to vary throughout the cycle. Enzymes that might play a rôle in the events of fertilization have been briefly commented upon by Aitken (1979) and by Mastroianni and Go (1979). Protease inhibitors were highlighted, and noted in the tubal fluids of rabbits, rhesus monkeys and women. Six protease inhibitors were identified in tubal fluid of rhesus monkeys and, whilst two of these were derived from serum, four were thought to stem from specific secretion. Alkaline phosphatase, amylase and lactate dehydrogenase have been identified in human tubal fluid (Pauerstein 1974), although β-amylase may be leaking from the epithelium (Edwards 1980). Diesterase (GPC) and lysozyme have also been noted in rabbit tubal fluid, and a broad spectrum of enzymes has been characterised in sheep fluid (Menezo 1975). In comparison with their concentrations in blood, many enzymes are found at lower levels in tubal fluid (Table III.2).

Amino acids present in the blood of rabbits are all revealed in the tubal fluid. Whereas most are present at reduced concentrations, those of glutamic acid and

Table III.2. A comparison of the relative activity of various enzymes in three different body fluids. The figures represent an arbitrary scale of values for enzyme activity extending from 0 to 3. (Adapted from Thibault 1972)

Enzymes	Activity		
	Tubal fluid	Blood serum	Seminal plasma
Alkaline phosphatase	2	2	2
Acid phosphatase	1	2	1
Esterases			
Acetate	3	3	3
Butyrate	2	3	2
Lipase	1	3	2
Aminopeptidase			
Alanyl	1	2	2
Leucyl	1	1	2
Valine	–	1	0
Galactosidase	0	0	1
Glucosaminidase	0	0	1
Glucuronidase	0	0	1
Glucosidase	0	0	1

glycine are apparently several times higher than in blood plasma (Thibault 1972). But, with respect to the oestrous cycle, serine, glutamic acid, alanine and glycine concentrations are all lower during oestrus and the pre-ovulatory period than during the beginning of the luteal phase in gonadotrophin-treated does (Menezo and Laviolette 1972). The extent to which these changes are related to the physiology of the gametes or embryos is uncertain, although the control of such amino acid concentrations in tubal fluid seems to reside in the changing patterns of secretion of gonadal steroid hormones. Amino acids essential for cleavage of the rabbit embryo in vitro have been reviewed by Thibault (1972): whilst the first cleavage can occur in a medium not supplemented with amino acids, the second cleavage requires cysteine, tryptophan, phenylalanine, lysine, arginine and valine. Subsequent division up to the morula stage also requires the addition of methionine, threonine and glutamine (Daniel and Olson 1968). Amino acids present in the tubal fluid of pigs and sheep have been summarized by Hamner (1973); glycine is the predominant amino acid in pig tubal fluid (Iritani et al. 1974) and also in sheep (Menezo 1975).

In women, Moghissi (1971) has reported cyclic variation in the amino acid concentration: when compared with serum values, amino acids in tubal fluid were higher during the proliferative phase of the menstrual cycle than during the luteal phase.

Electrolyte Content. Observations on the electrolyte concentration of tubal fluid have been made in rabbits, farm animals and primates. As might be expected, sodium is a major cation and chloride a major anion. Differences between blood plasma and tubal fluid can be demonstrated in ionic composition, especially after ovariectomy and/or steroid hormone therapy. The concentration of potassium in the tubal fluid of sheep, intact or ovariectomized, is several times higher than in blood, whereas calcium and magnesium are found in lower concentrations. However, although lower than in blood serum, the calcium content of rabbit, rhesus monkey and human tubal fluid increases after ovulation (Holmdahl and Mastroianni 1965; Mastroianni, Urzua and Stambaugh 1973; Lippes 1979). Phosphate and bicarbonate concentrations are below those of blood, although the bicarbonate level of tubal fluid is increased during oestrus or after oestrogen treatment of sheep (Restall and Wales 1966); this change in bicarbonate tension could have an important influence on respiration, capacitation and the consequential hyperactivation of spermatozoa (see Chap. V). The bicarbonate ion in rabbit tubal fluid exceeds the concentration in blood plasma and has been assigned responsibility for dispersing corona cells around freshly-shed oocytes (Stambaugh, Noriega and Mastroianni 1969), and bicarbonate ion in rhesus monkey tubal fluids may serve a similar purpose; its concentration also increases following ovulation (Maas, Storey and Mastroianni 1977).

Species differences in the electrolyte content of tubal fluid in response to gonadal steroid hormones have been summarized by Hamner (1973) and Coutinho (1974). Sheep tubal fluid concentrations of sodium, potassium and chloride are significantly lower at the end of oestrus than during the luteal phase of the cycle, whilst magnesium levels are lowest at oestrus. In the cow, by contrast, sodium and potassium concentrations in tubal fluid are lowest at oestrus whereas

Table III.3. Values for the concentration (mean ± S.E.M.) of various elements in the fluid of human Fallopian tubes before and after ovulation. (Taken from Borland et al. 1980)

Element	Hospital normal ranges for human serum	Tubal fluid (mM)				
		Borland et al. (1979) (N = 7)	Lippes et al. (1972) (N = 16)		David et al. (1973)	
			Pre-ovulatory	Post-ovulatory	Pre-ovulatory (N = 13)	Post-ovulatory (N = 11)
Na	139 – 147	130 ± 5	140 ± 3	139 ± 2	142 ± 4	148 ± 5
Cl	102 – 113	132 ± 6	119 ± 4	117 ± 3	127 ± 20	112 ± 11
K	3.6 – 5.0	21.2 ± 0.24	9.9 ± 1.8	7.7 ± 0.9	6.7 ± 1.3	6.7 ± 0.5
Ca	2.27 – 2.72	1.13 ± 0.24	1.89 ± 0.52	2.37 ± 0.27	–	–
Mg	0.55 – 1.57	1.42 ± 0.26	0.53 – 0.69	–	–	–
S	–	12.3 ± 1.9	–	–	–	–
P	–	8.69 ± 0.64	–	–	–	–

calcium levels are at a peak. Although these observations are of considerable interest, the reader should treat them with caution due to the different methods employed for fluid collection.

Human tubal fluid contains relatively high potassium and chloride concentrations but a low calcium concentration when compared with blood serum (Lippes et al. 1972; David et al. 1973; Borland et al. 1980). The sodium, magnesium and phosphorus concentrations in tubal fluid are similar to those in blood serum (Lippes et al. 1972; Borland et al. 1980), although the report of Lippes (1979) indicates lower values for magnesium and phosphorus in tubal fluid. Some differences in the composition of human tubal fluid before and after ovulation are presented in Table III.3. The physiological significance of the high potassium and low calcium concentrations in the events of fertilization and/or early embryonic development has yet to be determined, as has the manner in which these relative levels are generated via the mucosa; active transport across the epithelial cells would certainly seem to be involved.

Energy Substrates. A substantial range of energy substrates has been documented in tubal fluid. Amongst those most frequently highlighted in experimental animals are pyruvate, lactate, glucose and glycogen. Such compounds are thought to be involved in the metabolic support of the gametes and/or the pre-implantation embryos (Brinster 1973; Biggers and Borland 1976).

Fluid from the Fallopian tubes of oestrous farm animals or mid-cycle primates causes maximum stimulation of sperm metabolism. Several constituents, including glucose, pyruvate, lactate, and — as noted above — bicarbonate are responsible for such stimulation of respiration and glycolysis (Hamner 1973). The concentrations of glucose, pyruvate and lactate in rodent and rabbit tubal fluid all increase significantly after ovulation, perhaps inferring a critical rôle in development of the embryos (Holmdahl and Mastroianni 1965). Mean pyruvate concentrations increase from oestrous values of 16.1 µg/ml to 22.1 µg/ml after ovulation. Changes in lactate concentrations in rabbit tubal fluid are even more striking, and in vitro culture studies demonstrate lactate to be a principal energy source for developing zygotes. During progesterone therapy to mimic the luteal phase, the rabbit ampulla produces significantly more lactate than the isthmus (Hamner 1973). Concerning sources of pyruvate in tubal fluid, it should be borne in mind that granulosa cells from the mature follicle(s) will have entered the ampulla at ovulation and may have elaborated pyruvate from glucose (Biggers, Whittingham and Donahue 1967; Donahue and Stern 1968).

In fluid from cows with indwelling tubal cannulae, Carlson et al. (1970) observed a cyclic variation in glucose concentration in one animal, with highest values on the first day after oestrus. Pyruvate, lactate and glucose are also found in the tubal fluid of rhesus monkeys and women. The glucose level of human tubal fluid is about one third to one half that of serum (Lippes et al. 1972). As to an involvement of glycogen, granules of this energy substrate are released into the lumen of the mouse isthmus after ovulation, and 2-cell mouse embryos increase in glycogen content whilst in this portion of the Fallopian tube (Stern and Biggers 1968; Reinius 1970), supporting a functional rôle for this secretion.

Table III.4. Some constituents of fluid from the Fallopian tubes of sheep according to the stage of the oestrous cycle. (Summarized by Hamner 1973)

Constituent[a]	Hormonal state		
	Oestrus	Metoestrus	Dioestrus
Sodium, mg	3.2	3.0	3.2
Chloride, mg	4.6	4.4	4.8
Potassium, µg	320	295	317
Magnesium, µg	19.8	21.8	24.6
Calcium, µg	122	114	121
Phosphorus (PO_4^{3-}), µg	8.7	4.6	8.4
Total phosphorus, µg	40.5	21.3	46.3
Bicarbonate, mg	1.2	1.0	1.1
Glucose, µg	285 − 318	245 − 335	
Fructose, µg	0.52	0.89	
Pyruvate, µg	16.2	16.3	
Lactate, µg	150 − 226	114 − 422	382
Citrate, µg	8		
Lipid choline, µg	10	5	
Total protein, mg	6.7 − 30	4.8 − 28	2.1 − 32
Carbohydrate, µg	613	437	670
Dry matter, mg	32	40	

[a] Mean values expressed as unit/ml.

The details presented above are but examples of constituents of tubal fluid, and others are given in Table III.4 and in the review by Hamner (1973). It is not easy to state an unequivocal involvement for any single fluid constituent in the events of fertilization and early development of the embryo in vivo, not least because of the probability of highly complex interactions at the surface of (and within) the gametes or embryos. On the other hand, there is the preceding evidence of a change in the availability of metabolic substrates in tubal fluid, and − based largely on studies in laboratory rodents (Brinster 1973; Biggers and Borland 1976) − these changes match the embryos' requirement for and ability to utilize the substrates.

A novel advance in approaches to studying the composition of tubal fluid has been an in vivo rabbit preparation, in which the techniques of vascular and luminal perfusion are combined to examine transport processes across the epithelium (Leese and Gray 1985). Whilst this is an exciting short-term model for dynamic studies, events in an exposed and cannulated Fallopian tube bearing ligatures may not give a strict reflection of the physiological situation. Nonetheless, the observation that omission of glucose and pyruvate from the vascular perfusate resulted in barely detectable amounts in the lumen of the tube does suggest that these components are derived principally from the blood.

Physical Characteristics. The osmolality of tubal fluid from oestrous rabbits ranges from $302-310$ mOsmol (Stambaugh et al. 1969). Progesterone therapy of oestrogen-primed does and ovariectomy both increase osmolarity, although within a reduced volume of fluid; after ovariectomy, the fluid reaches an average osmolality of $371-388$ mOsmol (Hamner and Fox 1969).

The presumed pH within the Fallopian tubes has been recorded in a wide range of animal models, and the values are of the same order. Thus, collected samples of rabbit tubal fluid have a pH of $7.7-8.2$ in ligation experiments (Vishwakarma 1962) or of approximately 7.8 during oestrus, rising to $8.3-8.5$ during early pregnancy (Foley et al. 1972). Fluctuations in bicarbonate concentration in tubal fluid, induced by the action of ovarian steroid hormones, are an important cause of the change in pH and will be modified by CO_2 loss during sample collection. Avoiding such loss of CO_2, measurements in oestrous rabbits gave a pH value of 7.5 (Brackett and Mastroianni 1974). Within different regions of the rabbit Fallopian tube after microsurgical ligation, the pH varied from $7.3-8.2$ but without any obvious directional trend (David et al. 1969).

Fluid within the tubes of oestrous pigs has registered a pH value of 7.9 (Engle et al. 1968) or 8.1 just before ovulation on the second day of oestrus (author's unpublished observations). In rhesus monkeys, there is good evidence of an ovarian endocrine influence on pH with values of $7.1-7.3$ during the follicular phase of the menstrual cycle followed by a sudden increase to $7.5-7.8$ at the time of ovulation; pH remained high during the luteal phase of the cycle (Maas et al. 1977). pH values in women with hydrosalpinx range from $7.3-7.7$ (David et al. 1973).

Oxygen tensions within tubal fluid have been reported in rabbits by Bishop (1956, 1969) and in rhesus monkeys by Maas, Storey and Mastroianni (1976). The latter group has also recorded values for pCO_2 (Maas et al. 1977). Gas tensions will influence both the activity of spermatozoa and the developmental potential of the young embryo, and may well differ along the length of the tube.

Regional and Microenvironments

Preliminary studies on tubal fluid analysed the contents or flushings of the duct lumen in a somewhat gross manner, the accumulated fluids apparently being regarded as homogeneous. But, by the early 1970s, there was a growing appreciation that the different regions of the tube — the infundibulum, ampulla and isthmus — might each provide a differing fluid environment for the gametes or embryos (see Hamner 1973; Freese, Orman and Paulos 1973). By the late 1970s, largely due to the sophistication of electron probe analysis used by Biggers and his colleagues, it was possible to go beyond simple regional analysis of tubal fluids and demonstrate that there were indeed microenvironments, especially in the vicinity of developing embryos. This observation suggested a local influence of embryonic signals upon transudation and/or secretion by the tubal epithelium.

Preliminary studies had noted that after placing ligatures on the Fallopian tubes, fluid accumulation in the lumen was several times more extensive in the ampulla than the isthmus (Hamner 1973). Whilst this might merely have reflected the distensibility of the muscular coats, it was also possible that the activity of the epithelial layers varied regionally in relation to the processes of transudation and secretion. Gradients in the concentration of tubal fluid constituents would then be anticipated on the basis of a dilution effect. Further circumstantial evidence for regional differences came from studies in mice, in which the structure and histochemistry of the tubal epithelium varied from the infundibulum to the utero-tubal junction (Reinius 1970). Secretory material, thought to be composed of mucoproteins and acid mucopolysaccharides, is released into the ampullary lumen after ovulation, and the fluid in this region bathes the eggs and distends the ampulla. Similarly, cleaving embryos can subsequently be demonstrated within discrete, fluid-filled loops of the isthmus, whilst other portions of the isthmus show no such distension (Reinius 1970).

Another of the early approaches to demonstrating regional differences in the composition of luminal fluid involved a microsurgical preparation in which the rabbit Fallopian tube was separated into four segments, with the absolute minimum of disturbance to the blood supply (David et al. 1969). The volume of fluid collected from the ligated segments decreased from the ovarian to the uterine end, and chemical constituents varied in concentration within fluids from the different segments. Whilst the concentration of chloride decreased significantly in the isthmic segment, the concentrations of sodium, bicarbonate, inorganic phosphate, proteins and lactic acid increased significantly from the ovarian end of the tube towards the uterus. These results supported the possibility that different segments of the tube might play specific rôles vis à vis the gametes or embryos.

Gupta, Karkun and Kar (1970) were the first specifically to appreciate that the presence of a rabbit embryo in the tubal lumen might influence epithelial synthesis and hence the composition of tubal tissues. They found that the ampulla had a higher lactate concentration when fertilized eggs were present in this portion of the tube, and the same applied to the ampullary-isthmic junction and likewise to the isthmus, perhaps corresponding to the increased lactate in tubal fluid noted above. A local influence of the embryo on the synthetic activity of the tubal epithelium was again raised by Stone and Hamner (1975) and Hunter (1977) but the evidence was slender. Intuition, presumably, prompted Stone and Hamner (1975) to remark that "... certain tubal secretions, possibly released only by epithelial contact with a living zygote, may prove essential in directing embryological differentiation".

In a major review, Biggers and Borland (1976) stressed the fact that − at any given time − the embryo is surrounded by a microenvironment in the Fallopian tube with which it interacts by the exchange of substances. Using the technique of electron probe microanalysis, Roblero, Biggers and Lechene (1976) reported that the elemental composition of ampullary and isthmic fluids bathing mouse embryos was dissimilar, and that both were very different in composition from blood serum (Table III.5). The mean potassium, sulphur and phosphorus contents of fluid obtained from the isthmus containing 2-cell embryos were significantly higher than those of the ampullary fluid containing recently-fertilized eggs and follicular cells. These observations on undisturbed samples of tubal fluid were the first to present chemical analyses of the microenvironment of the preimplantation mammalian embryo, the only reservation being that they stemmed

44

Table III.5. The weighted mean concentrations (mM/l) of various elements in serum and the microenvironments of 1- and 2-celled mouse embryos in the ampulla and isthmus, respectively, of the oviduct. (Adapted from Roblero et al. 1976)

Element	Ampulla (N = 25)	Isthmus (N = 17)	Serum (N = 12)
Na	137 ± 4.78	142 ± 3.42	148 ± 1.72
Cl	147 ± 4.98	145 ± 6.20	140 ± 2.10
Ca	1.62 ± 0.13	1.48 ± 0.14	3.82 ± 0.16
K	17.8 ± 1.38	29.7 ± 3.63^a	5.26 ± 0.15
Mg	0.63 ± 0.08	0.81 ± 0.07	1.38 ± 0.20
S	5.13 ± 0.53	9.13 ± 0.92^a	21.8 ± 0.32
P	3.90 ± 0.64	8.46 ± 1.79^a	6.97 ± 0.27

[a] Significantly different ($P < 0.01$) from the concentration in ampullary fluid (Fisher-Behren's test). All other comparisons are not statistically significant at the $P = 0.05$ level.

from gonadotrophin-treated immature mice. In a subsequent paper from the same group using mature mice (Borland et al. 1977), the increase in potassium and phosphorus concentrations from the ampulla to the isthmus was not endorsed. No satisfactory explanation for the discrepancy was provided. The authors went on to record differences between bursal sac fluid and ampullary and isthmic fluids, emphasizing in the latter the high concentrations of potassium (23 – 25 mM) and low concentrations of calcium (1.7 – 2.0 mM); the high levels of potassium were thought to originate from the tubal epithelium.

It would be valuable to demonstrate specific changes in the vascular bed and rate of blood and lymph flow in the tubal tissues surrounding the embryos, but such an advance will almost certainly require the application of sophisticated new microtechnology. Use of sheep or pigs as experimental animals rather than rats or mice would have several advantages in this kind of work. The accumulating evidence for one or more early pregnancy factors from the very young zygote (see Chap. VI) would seem to point even more persuasively to local influences of the embryo upon the tubal epithelium, and thus to the likelihood of quite distinct microenvironments during progression along the isthmus to the uterus.

Hormones Detected in Tubal Fluids

Because hormones are widely distributed in body fluids, it is not surprising that they are detectable within the lumen of the Fallopian tubes; this is especially so when the contribution of blood transudation to tubal fluids is recalled. Most of the measurements to date concern steroid hormones and prostaglandins, but there is tentative evidence also for prolactin, gonadotrophins and various peptide hormones in tubal fluid. Binding sites for oxytocin and relaxin in the epithelial tissues of the Fallopian tube might indicate that these hormones act in part via the luminal fluids. Difficult questions to resolve satisfactorily are (1) whether hormones in tubal fluid have origins quite distinct from the blood plasma, and (2) whether such hormones have a specific involvement in the physiology of the gametes and development of the embryos.

Free steroid hormones (i.e. unbound) have been detected in the tubal fluid of rhesus monkeys (Wu, Mastroianni and Mikhail 1977). Hormone concentrations were lower than those in blood plasma before ovulation, whilst oestrone and testosterone rose to or above plasma values during the post-ovulatory period. After ovulation, there was a gradual increase in the tubal fluid concentrations of oestrone, testosterone and progesterone, whereas the concentrations of oestradiol and androstenedione decreased. The enhanced steroid concentrations in tubal fluid might involve epithelial conversion of precursor substrates or, alternatively, enzymatic liberation of protein-bound hormones.

In oestrous and pseudopregnant rabbits, the concentrations of oestradiol and progesterone in the tubal fluid were first reported by Richardson and Oliphant (1981). Oestradiol values were similar to those in serum (48−120 pg/ml) whereas progesterone concentrations ranged from 0.6−2.9 ng/ml; progesterone values were lowest at oestrus and maximum on Day 12 of pseudopregnancy. A divergence developed between the concentration of progesterone in tubal fluid and that in serum as pseudopregnancy advanced, serum values becoming significantly higher.

The source of steroid (and other) hormones in tubal fluids is open to debate. Whilst it is probable that these small molecules are derived principally from a blood transudate, contributions from peritoneal fluid, ovarian follicular fluid, granulosa cells and from the embryos themselves cannot be overlooked. The antral fluid of mature Graafian follicles is rich in a variety of steroid hormones and, although this gonadal fluid is thought not to pass down the ampulla into the isthmus − due to the retrograde direction of flow at this time (see above) − it could still influence the composition of native tubal fluid. On the other hand, the granulosa cells released at ovulation do progress along the Fallopian tube to the uterus, at least in some species (Hunter 1978), and these cells may well continue to synthesize steroids and peptide hormones and modify the immediate environment of the embryo(s) and/or tubal epithelium. In that granulosa cells show steroid synthetic activity during in vitro culture with suitable substrates, it is highly probable that they can function similarly in vivo. As to a hormonal contribution from the embryos themselves, there are several reports that they contain hydroxysteroid dehydrogenase activity as early as 48 h post coitum in the rabbit and hamster (Dickmann, Dey and Gupta 1975; Niimura and Ishida 1976), so it is possible that the embryo makes an active contribution to the endocrine environment within the Fallopian tube.

Turning to prostaglandins, although numerous studies have documented the presence of these hormones in the wall of the Fallopian tube (e.g. Ogra et al. 1974; Saksena and Harper 1975; Vastik-Fernandez et al. 1975; Oyamada 1981; Thomas, Bastiaans and Rolland 1982), relatively few studies have determined their concentration in the luminal fluid. But, by analogy with the potential of the uterus to secrete prostaglandins in an exocrine manner (Bazer and Thatcher 1977), it seemed reasonable to anticipate an exocrine secretion into the tubal lumen.

Fluid from sheep prepared with indwelling cannulae has been reported to contain remarkably high levels of prostaglandins of the F series (PGF), values reaching 230 ng/ml (Warnes, Amato and Seamark 1978). However, concern was expressed in this report that the cannulae may have caused irritation of the tubal

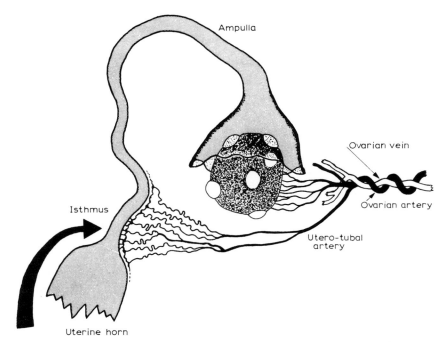

Fig. III.5. A semi-diagrammatic representation of the arterial blood supply to the ovary and isthmus of the pig Fallopian tube. A portion of the ovarian vein is also shown. A counter-current transfer of follicular hormones was demonstrated from the ovarian vein to the corresponding artery and thus into the utero-tubal branch (Hunter et al. 1983)

epithelium, thereby leading to unphysiological levels of prostaglandins in the fluid. In women, the concentration of $PGF_{2\alpha}$ in tubal fluid shows cyclic variation, with higher values in pre-ovulatory compared with post-ovulatory fluids (7.1 vs. 5.0 ng/ml, respectively, Lippes 1979). The concentration of unbound $PGF_{2\alpha}$ in the tubal fluid at the time of ovulation was much higher than in the systemic circulation (Ogra et al. 1974), but this would be a general expectation as a result of metabolic clearance in the pulmonary circulation. Whilst Ogra et al. (1974) suggested synthesis of $PGF_{2\alpha}$ by the tubal mucosa, an ovarian contribution from the high titres within mature Graafian follicles should not be overlooked. A counter-current transfer system from the ovarian vein to the tubal branch of the ovarian artery might well be operational for this hormone (see Hunter, Cook and Poyser 1983; Einer-Jensen, Bendz and Leidenberger 1985; Fig. III.5). Such a means of transfer might also exist for oxytocin and relaxin (see Schramm, Einer-Jensen and Schramm 1986; Schramm et al. 1986), which are found in the pre-ovulatory follicle.

Because of the fact that a prolactin release from the endometrium has been reported during the menstrual cycle in women (Maslar and Riddick 1979) and is demonstrable in uterine fluid or flushings (Stone et al. 1986), there is a good possibility that prolactin will also be demonstrable in tubal fluid; this is not least because of the structure of the intra-mural portion of the tube, permitting access of uterine fluid when muscle tone is reduced. Specific reports from non-pathological specimens are still awaited, and this comment is also true for the peptide hormones oxytocin and relaxin.

As to a rôle for hormones in tubal fluid, steroids have been widely reported to influence spermatozoa directly (e.g. Ericsson, Cornette and Buthala 1967) and likewise to influence developing embryos, at least during in vitro culture (e.g. McGaughey and Daniel 1966; Edwards 1980). Of considerable significance may be the observation in vivo that monoclonal antibodies against progesterone administered by the intraperitoneal or intravenous routes inhibit cleavage of mouse embryos in the Fallopian tube (Wright et al. 1982; Wang et al. 1984; Heap et al. 1986). The ability of a specific anti-progesterone antibody to block cleavage by passive immunization is thought to be associated with modified tubal secretions, but a component of such arrested development might include a direct influence of the antibody on the embryo itself, even though this is apparently not demonstrable by means of in vitro culture (Heap et al. 1986); however, subtle molecular interactions might underlie a putative in vivo effect. The most recent publication from this group indicated embryonic arrest in vivo shortly after the 4-cell stage, and presents evidence that the sensitivity of the underlying mechanisms to passive immunization against progesterone is influenced by the genotype of the treated mice (Rider et al. 1987).

A different focus of involvement for both steroid hormones and prostaglandins is the one of programming the amplitude and frequency of myosalpingeal contractions, possibly via the regulation of calcium transport. These remarks also apply to the frequency and directional beat of the cilia (see Chaps. IV and VII).

As a concluding remark, the time would seem ripe for a further investigation of the biochemistry of the fluids found in the lumen of the Fallopian tubes, and especially for further study of the microenvironments surrounding developing embryos. The methodology would need to be sophisticated: with the notable exception of the electron probe studies referred to above, much of the work so far seems to have lacked imaginative sparkle and specific orientation. If the composition of the microenvironment around the eggs could be shown to differ, for example, between secondary oocytes and developing embryos, then this would be of special interest in terms of early pregnancy factors. And, in the context of formation of fluids in the Fallopian tubes, comparison of the physiology with that underlying formation of ovarian follicular fluid might be revealing; the transporting nature of the endosalpinx and that of the follicular epithelium may well have striking features in common. Both provide fluids that bathe the oocyte.

REFERENCES

Aitken RJ (1979) Tubal and uterine secretions: the possibilities for contraceptive attack. J Reprod Fertil 55:247–254
Bazer FW, Thatcher WW (1977) Theory of maternal recognition of pregnancy in swine based on estrogen controlled endocrine versus exocrine secretion of prostaglandin $F_{2\alpha}$ by the uterine endometrium. Prostaglandins 14:397–401

Bazer FW, Roberts M, Sharp DC (1978) Collection and analysis of female genital tract secretions. In: Daniel JC Jr (ed) Methods in mammalian reproduction. Academic Press, New York, pp 503–528

Beier HM (1974) Oviducal and uterine fluids. J Reprod Fertil 37:221–237

Bellve AR, McDonald MF (1968) Directional flow of Fallopian tube secretion in the Romney ewe. J Reprod Fertil 15:357–364

Bellve AR, McDonald MF (1970) Directional flow of Fallopian tube secretion in the ewe at onset of the breeding season. J Reprod Fertil 22:147–149

Biggers JD, Borland RM (1976) Physiological aspects of growth and development of the preimplantation mammalian embryo. Ann Rev Physiol 38:95–119

Biggers JD, Whittingham DG, Donahue RP (1967) The pattern of energy metabolism in the mouse oocyte and zygote. Proc Natl Acad Sci USA 58:560–567

Bishop DW (1956) Active secretion in the rabbit oviduct. Am J Physiol 187:347–352

Bishop DW (1969) Sperm physiology in relation to the oviduct. In: Hafez ESE, Blandau RJ (eds) The mammalian oviduct. University of Chicago Press, Chicago, pp 231–250

Björkman N, Fredricsson B (1966) The bovine oviduct epithelium and its secretory process as studied with the electron microscope and histochemical tests. Z Zellforsch Mikrosk Anat 55:500–513

Black DL, Asdell SA (1959) Mechanism controlling entry of ova into rabbit uterus. Am J Physiol 197:1275–1278

Black DL, Davis J (1962) A blocking mechanism in the cow oviduct. J Reprod Fertil 4:21–26

Borland RM, Hazra S, Biggers JD, Lechene CP (1977) The elemental composition of the environments of the gametes and preimplantation embryo during the initiation of pregnancy. Biol Reprod 16:147–157

Borland RM, Biggers JD, Lechene CP, Taymor ML (1980) Elemental composition of fluid in the human Fallopian tube. J Reprod Fertil 58:479–482

Brackett BG, Mastroianni L (1974) Composition of oviducal fluid. In: Johnson AD, Foley CW (eds) The oviduct and its functions. Academic Press, New York, pp 133–159

Braden AWH (1952) Properties of the membranes of rat and rabbit eggs. Aust J Sci Res 5:460–471

Brinster RL (1973) Nutrition and metabolism of the ovum, zygote and blastocyst. In: Greep RO, Astwood EB (eds) Handbook of Physiology, Section 7, Endocrinology II, American Physiological Society, Washington DC, pp 165–185

Carlson D, Black DL, Howe GR (1970) Oviduct secretion in the cow. J Reprod Fertil 22:549–552

Clewe TH, Mastroianni L (1960) A method for continuous volumetric collection of oviduct secretions. J Reprod Fertil 1:146–150

Coutinho EM (1974) Hormonal control of oviductal motility and secretory functions. In: Greep RO (ed) Reproductive physiology. MTP Intl Rev Sci. Butterworths, London, pp 133–153

Daniel JC, Olson JD (1968) Amino acid requirement for cleavage of the rabbit ovum. J Reprod Fertil 15:453–456

David A, Brackett BG, Garcia CR, Mastroianni L (1969) Composition of rabbit oviduct fluid in ligated segments of the Fallopian tube. J Reprod Fertil 19:285–289

David A, Serr DM, Czernobilsky B (1973) Chemical composition of human oviduct fluid. Fertil Steril 24:435–439

Dickmann Z, Dey SK, Gupta JS (1975) Steroidogenesis in rabbit preimplantation embryos. Proc Natl Acad Sci USA 72:298–300

Donahue RP, Stern S (1968) Follicular cell support of oocyte maturation: production of pyruvate in vitro. J Reprod Fertil 17:395–398

Edwards RG (1980) Conception in the human female. Academic Press, London

Einer-Jensen N, Bendz A, Leidenberger F (1985) Transfer of 13-C progesterone in the human ovarian adnex. Proc Soc Stud Fertil Abstract no 53:41

Engle CC, Dunn JS, Hood RO, Williams DJ, Foley CW, Trout HF (1968) Amino acids in sow and rabbit oviduct fluids. J Anim Sci 27:1786

Ericsson RJ, Cornette JC, Buthala DA (1967) Binding of sex steroids to rabbit sperm. Acta Endocrinol Copenh 56:424–432

Fazleabas AT, Verhage HG (1986) The detection of oviduct-specific proteins in the baboon (*Papio anubis*). Biol Reprod 35:455–462

Feigelson M, Kay E (1972) Protein patterns of rabbit oviducal fluid. Biol Reprod 6:244–252

Foley CW, Engle CC, Plotka ED, Roberts RC, Johnson AD (1972) Influence of early pregnancy on proteins and amino acids of uterine tubal fluids in rabbits. Am J Vet Res 33:2059–2065

Fredricsson B (1969) Histochemistry of the oviduct. In: Hafez ESE, Blandau RJ (eds) The mammalian oviduct. University of Chicago Press, Chicago, pp 311–312

Freese UE, Orman S, Paulos G (1973) An autoradiographic investigation of epithelium – egg interaction in the mouse oviduct. Am J Obstet Gynecol 117:364–370

Glass LE (1969) Immunocytological studies of the mouse oviduct. In: Hafez ESE, Blandau RJ (eds) The mammalian oviduct. University of Chicago Press, Chicago, pp 459–476

Glass LE, McClure TR (1965) Postnatal development of the mouse oviduct: transfer of serum antigens to the tubal epithelium. In: Wolstenholme GEW, O'Connor M (eds) Preimplantation stages of pregnancy. CIBA Foundation Symposium. Churchill, London, pp 294–321

Gupta DN, Karkun J, Kar AB (1970) Biochemical changes in different parts of the rabbit Fallopian tube during passage of ova. Am J Obstet Gynecol 106:833–837

Hamner CE (1973) Oviducal fluid – composition and physiology. In: Greep RO, Astwood EB (eds) Handbook of physiology, Section 7, Endocrinology II. American Physiological Society, Washington DC, pp 141–151

Hamner CE, Fox SB (1969) Biochemistry of oviductal secretions. In: Hafez ESE, Blandau RJ (eds) The mammalian oviduct. University of Chicago Press, Chicago, pp 333–355

Hamner CE, Williams WL (1965) Composition of rabbit oviduct secretions. Fertil Steril 16:170–176

Heap RB, Rider V, Wooding FBP, Flint APF (1986) Molecular and cellular signalling and embryo survival. In: Sreenan JM, Diskin MG (eds) Embryonic mortality in farm animals. Nijhoff, Dordrecht, pp 46–73

Holmdahl T, Mastroianni L Jr (1965) Continuous collection of rabbit oviduct secretions at low temperature. Fertil Steril 16:587–595

Hunter RHF (1977) Function and malfunction of the Fallopian tubes in relation to gametes, embryos and hormones. Europ J Obstet Gynecol Reprod Biol 7:267–283

Hunter RHF (1978) Intraperitoneal insemination, sperm transport and capacitation in the pig. Anim Reprod Sci 1:167–179

Hunter RHF (1980) Physiology and technology of reproduction in female domestic animals. Academic Press, London

Hunter RHF, Cook B, Poyser NL (1983) Regulation of oviduct function in pigs by local transfer of ovarian steroids and prostaglandins: a mechanism to influence sperm transport. Europ J Obstet Gynecol Reprod Biol 14:225–232

Hyde BA, Black DL (1986) Synthesis and secretion of sulphated glycoproteins by rabbit oviduct explants in vitro. J Reprod Fertil 78:83–91

Iritani A, Sato E, Nishikawa Y (1974) Secretion rates and chemical composition of oviduct and uterine fluids in sows. J Anim Sci 39:582–588

Johnson AD, Foley CW (1974) The oviduct and its function. Academic Press, New York

Kapur RP, Johnson LV (1985) An oviductal fluid glycoprotein associated with ovulated mouse ova and early embryos. Develop Biol 112:89–93

Leese HJ, Gray SM (1985) Vascular perfusion: a novel means of studying oviduct function. Am J Physiol 248:E624–E632

Lippes J (1975) Applied physiology of the uterine tube. Obstet Gynecol Ann 4:119–166

Lippes J (1979) Analysis of human oviductal fluid for low molecular weight compounds. In: Beller FK, Schumacher GFB (eds) The biology of the fluids in the female genital tract. Elsevier, North Holland, pp 373–387

Lippes J, Enders RG, Pragay DA, Bartholomew WR (1972) The collection and analysis of human Fallopian tubal fluid. Contraception 5:85–103

Lippes J, Krasner J, Alfonso LA, Dacalos ED, Lucero R (1981) Human oviductal fluid proteins. Fertil Steril 36:623–629

Maas DHA, Storey BT, Mastroianni L (1976) Oxygen tension in the oviduct of the rhesus monkey (*Macaca mulatta*). Fertil Steril 27:1312–1317

Maas DHA, Storey BT, Mastroianni L Jr (1977) Hydrogen ion and carbon dioxide content of the oviductal fluid of the rhesus monkey (*Macaca mulatta*). Fertil Steril 28:981−985

Maslar IA, Riddick DH (1979) Prolactin production by human endometrium during the menstrual cycle. Am J Obstet Gynecol 135:751−754

Mastroianni L Jr, Go KJ (1979) Tubal secretions. In: Beller FK, Schumacher GFB (eds) The biology of the fluids of the female genital tract. Elsevier, North Holland, pp 335−344

Mastroianni L, Shah U, Abdul-Karim R (1961) Prolonged volumetric collection of oviduct fluid in Rhesus monkey. Fertil Steril 12:417−424

Mastroianni L, Urzua MA, Stambaugh R (1970) Protein patterns in monkey oviducal fluid before and after ovulation. Fertil Steril 21:817−820

Mastroianni L Jr, Urzua M, Stambaugh R (1973) The internal environmental fluids of the oviduct. In: Segal SJ, Crozier R, Corfman PA, Condliffe PG (eds) The regulation of mammalian reproduction. Thomas, Springfield, pp 376−384

McDonald MF, Bellve AR (1969) Influence of oestrogen and progesterone on flow of fluid from the Fallopian tube in the ovariectomized ewe. J Reprod Fertil 20:51−61

McGaughey RW, Daniel JC (1966) Effect of oestradiol-17β on fertilized rabbit eggs in vitro. J Reprod Fertil 11:325−331

Menezo YJR (1975) Amino constituents of tubal and uterine fluids of the estrous ewe: comparison with blood serum and ram seminal fluid. In: Hafez ESE, Thibault CG (eds) The biology of spermatozoa. Karger, Basel, pp 174−181

Menezo YJR (1979) La composition chimique des sécrétions tubaires et ses variations: rôle dans la réalisation de la fécondation et les débuts de l'embryogenèse. In: Brosens I et al (eds) Oviducté et Fertilité, Proc SNESF, Masson, Paris, pp 109−126

Menezo YJR, Laviolette P (1972) Les constituants aminés des sécrétions tubaires chez la lapine. Annls Biol Anim Biochim Biophys 12:383−396

Moghissi KS (1970) Human Fallopian tube fluid. I. Protein composition. Fertil Steril 21:821−829

Moghissi KS (1971) Proteins and amino acids of human tubal fluid. In: Kleinman R, Pickles VR (ed) Excerpta medica. Int Congr Ser 234a. Excerpta Medica, Amsterdam

Murray FA, Goode L, Linnerud AC (1969) Effects of season, mating and pregnancy on the volume and protein content of ewe oviduct fluid. J Anim Sci 29:727−733

Niimura S, Ishida K (1976) Histochemical studies of Δ^5-3β, 20a- and 20β-hydroxysteroid dehydrogenases and possible progestagen production in hamster eggs. J Reprod Fertil 48:275−278

Ogra SS, Kirton KT, Tomasi TB, Lippes J (1974) Prostaglandins in the human Fallopian tube. Fertil Steril 25:250−255

Oliphant G, Reynolds AB, Smith PF, Ross PR, Marta JS (1984) Immunocytochemical localization and determination of hormone-induced synthesis of the sulfated oviductal glycoproteins. Biol Reprod 31:165−174

Oyamada T (1981) Levels of prostaglandin F_{2a} and E_2 in the human uterine tissues and in human Fallopian tubes from non-pregnant women. Acta Obstet Gynaecol Jpn 33:155−160

Parker GH (1931) The passage of sperms and of eggs through the oviducts in terrestrial vertebrates. Phil Trans 219:381−419

Pauerstein CJ (1974) The Fallopian tube: a reappraisal. Lea & Febiger, Philadelphia

Perkins JL, Goode L (1966) Effects of stage of the oestrous cycle and exogenous hormones upon the volume and composition of oviduct fluid in ewes. J Anim Sci 25:465−471

Perkins JL, Goode L, Wilder WA, Henson DB (1965) Collection of secretions from the oviduct and uterus of the ewe. J Anim Sci 24:383−387

Reinius S (1970) Morphology of oviduct, gametes and zygotes as a basis of oviductal function in the mouse. I. Secretory activity of the oviductal epithelium. Int J Fertil 15:191−209

Restall BJ (1966) The Fallopian tube of the sheep. II. The influence of progesterone and oestrogen on the secretory activities of the Fallopian tube. Aust J Biol Sci 19:187−197

Restall BJ, Wales RG (1966) The Fallopian tube of the sheep. III. The chemical composition of the fluid from the Fallopian tube. Aust J Biol Sci 19:687−698

Richardson LL, Oliphant G (1981) Steroid concentrations in rabbit oviducal fluid during oestrus and pseudopregnancy. J Reprod Fertil 62:427−431

Rider V, Heap RB, Wang MY, Feinstein A (1987) Anti-progesterone monoclonal antibody affects early cleavage and implantation in the mouse by mechanisms that are influenced by genotype. J Reprod Fertil 79:33–43

Roberts GP, Parker JM, Symonds HW (1975) Proteins in the luminal fluid from the bovine oviduct. J Reprod Fertil 45:301–313

Roberts GP, Parker JM, Symonds HW (1976) Macromolecular components of genital tract fluids from the sheep. J Reprod Fertil 48:99–107

Roblero L, Biggers JD, Lechene CP (1976) Electron probe analysis of the elemental microenvironment of oviducal mouse embryos. J Reprod Fertil 46:431–434

Saksena SK, Harper MJK (1975) Relationship between concentration of prostaglandin F (PGF) in the oviduct and egg transport in rabbits. Biol Reprod 13:68–76

Schramm W, Einer-Jensen N, Schramm G (1986) Direct venous-arterial transfer of ^{125}I-radiolabelled relaxin and tyrosine in the ovarian pedicle of sheep. J Reprod Fertil 77:513–521

Schramm W, Einer-Jensen N, Schramm G, McCracken JA (1986) Local exchange of oxytocin from the ovarian vein to ovarian arteries in sheep. Biol Reprod 34:671–680

Schumacher GFB, Yang SL, Broer KH (1979) Specific antibodies in oviduct secretions of the Rhesus monkey. In: Beller FK, Schumacher GFB (eds) The biology of the fluids of the female genital tract. Elsevier, North-Holland, pp 389–398

Shapiro SS, Brown NE, Yard AS (1974) Isolation of an acidic glycoprotein from rabbit oviducal fluid and its association with the egg coating. J Reprod Fertil 40:281–290

Stambaugh R, Noriega C, Mastroianni L Jr (1969) Bicarbonate ion; the corona cell dispersing factor of rabbit tubal fluid. J Reprod Fertil 18:51–58

Stanke DF, de Young DW, Sikes JD, Mather EC (1973) Collection of bovine oviduct secretions. J Reprod Fertil 32:535–537

Stanke DF, Sikes JD, de Young DW, Tumbleson ME (1974) Proteins and amino acids in bovine oviducal fluid. J Reprod Fertil 38:493–496

Stern S, Biggers JD (1968) Enzymatic estimation of glycogen in the cleaving mouse embryo. J Exp Zool 168:61–66

Stone BA, Petrucco OM, Seamark RF, Godfrey BM (1986) Concentrations of steroid hormones, and of prolactin, in washings of the human uterus during the menstrual cycle. J Reprod Fertil 78:21–25

Stone SL, Hamner CE (1975) Biochemistry and physiology of oviductal secretions. Gynecol Invest 6:234–252

Stone SL, Huckle WR, Oliphant G (1980) Identification and hormonal control of reproductive-tract-specific antigens in rabbit oviductal fluid. Gamete Res 3:169–177

Sutton R, Nancarrow CD, Wallace ALC, Rigby NW (1984) Identification of an oestrus-associated glycoprotein in oviducal fluid of sheep. J Reprod Fertil 72:415–422

Sutton R, Nancarrow CD, Wallace ALC (1986) Oestrogen and seasonal effects on the production of an oestrus-associated glycoprotein in oviducal fluid of sheep. J Reprod Fertil 77:645–653

Thibault C (1972) Physiology and physiopathology of the Fallopian tube. Int J Fertil 17:1–13

Thomas CMG, Bastiaans LA, Rolland R (1982) Concentrations of unconjugated oestradiol and progesterone in blood plasma and prostaglandins $F_{2\alpha}$ and E_2 in oviducts of hamsters during the oestrous cycle and in early pregnancy. J Reprod Fertil 66:469–474

Vastik-Fernandez J, Gimeno MF, Lima F, Gimeno AL (1975) Spontaneous motility and distribution of prostaglandins in different segments of human Fallopian tubes. Amer J Obstet Gynecol 122:663–668

Vishwakarma P (1962) The pH and bicarbonate-ion content of the oviduct and uterine fluids. Fertil Steril 13:481–485

Wang M-Y, Rider V, Heap RB, Feinstein A (1984) Action of anti-progesterone monoclonal antibody in blocking pregnancy after postcoital administration in mice. J Endocrinol 101:95–100

Warnes GM, Amato F, Seamark RF (1978) Prostaglandin F in the Fallopian tube secretions of the ewe. Aust J Biol Sci 31:275–282

Woskressensky MA (1891) Experimentelle Untersuchungen über die Pyo- und Hydrosalpinxbildung bei den Thieren. Zentralbl Gynäk 15:849–860

Wright LJ, Feinstein A, Heap RB, Saunders JC, Bennett RC, Wang M-Y (1982) Progesterone monoclonal antibody blocks pregnancy in mice. Nature (Lond) 295:415–417

Wu CH, Mastroianni L, Mikhail G (1977) Steroid hormones in monkey oviductal fluid. Fertil Steril 28:1250–1256

Yoshinaga K, Mahoney WA, Pincus G (1971) Collection of oviduct fluid from unrestricted monkeys. J Reprod Fertil 25:117–120

Transport of Gametes, Selection of Spermatozoa and Gamete Lifespans in the Female Tract

CONTENTS

Introduction   53
Shedding and Initial Transport of Eggs   53
Deposition and Transport of Spermatozoa.   58
Pre- and Peri-Ovulatory Sperm Distribution Within the Isthmus   66
Intra-Peritoneal Insemination   70
Sperm Selection in the Female Tract   71
Gamete Lifespans in the Female Tract   72
References   74

Introduction

The meeting of male and female gametes at the ampullary-isthmic junction of the Fallopian tube, the usual site of fertilization, has attracted much attention, not least from the point of view of the temporal relationships underlying this critical process and the numbers of viable spermatozoa in the vicinity of the egg(s) at the time of activation. In farm and laboratory animals in which there is a clear relationship between the onset of receptivity to the male (the period of oestrus) and the subsequent time of ovulation, transport of eggs and spermatozoa to the site of fertilization is thought to involve carefully regulated processes, for the eggs are usually penetrated by spermatozoa shortly after release from the ovary into the reproductive tract. In the biological situation, therefore, a mechanism seems to have evolved for avoiding post-ovulatory ageing of the egg – a condition known to have deleterious consequences. In some contrast, the timing of gamete entry into the Fallopian tubes of primates, including women, may be widely asynchronous due to the absence of any specific phase of oestrus. It is also worth noting that procedures of artificial insemination in farm animals may lead to asynchrony in the meeting of the gametes and thus to a lowered fertility when compared with the results of spontaneous mating.

Shedding and Initial Transport of Eggs

In the paragraphs that follow, transport of eggs refers principally to their displacement from the ovarian surface to the site of fertilization in the Fallopian tube. However, a brief discussion of the manner of expulsion of the oocyte from

Table IV.1. Examples of the approximate interval between the pre-ovulatory surge of gonadotrophic hormones detectable in the systemic circulation and collapse of the Graafian follicle(s) at ovulation

Species	Interval in hours
Rabbit	10
Sheep	25 – 26
Cow	28 – 30
Human	36 – 40
Pig	40 – 42

the collapsing follicle needs to be included, not least since this may suggest reasons underlying the occasional failure of egg release in primates, and thus the potential for initiating an ectopic (ovarian) pregnancy.

The interval between the pre-ovulatory gonadotrophin surge and ovulation, measured in hours (Table IV.1), permits a series of vital changes in the somatic cells and oocyte of the mature Graafian follicle(s) and in the fluid environment of steroid hormones, prostaglandins and diverse peptides. Relevant to the present discussion is the gradual separation of the granulosa cells, especially in the last hours before ovulation. The granulosa investment of the oocyte, the cumulus oophorus, itself undergoes a characteristic expansion and mucification with a rearrangement of cells surrounding the zona pellucida, the corona radiata (Fig. IV.1), and an elongation of their cytoplasmic processes (Dekel and Phillips 1979; Eppig 1980, 1982; Szöllösi and Gérard 1983; Bomsel-Helmreich 1985). These modifications in the somatic cells encourage separation of the oocyte from the follicle wall shortly before ovulation, although tenuous filamentous bridges may still act to anchor it loosely and on occasions prevent its release. Escape of the follicular fluid at ovulation is thought to aid displacement of the oocyte from the follicle (Blandau 1969, 1973; Fig. IV.2), and this process may be assisted by contractions in the thecal layers, perhaps promoted by locally synthesized prostaglandins (Le Maire et al. 1973; Bauminger and Lindner 1975). A rôle for the autonomic nervous system has been invoked to explain movements of the follicular wall at ovulation (O'Shea and Phillips 1974; Owman et al. 1975), for non-vascular adrenergic nerves are detectable in the wall of maturing follicles. The sympathetic nerve endings focus on actomyosin-containing cells in the theca externa (Amsterdam, Lindner and Gröschel-Stewart 1977), and electrical stimulation of these adrenergic fibres can induce contraction of the follicle wall (Walles, Owman and Sjöberg 1982). On the other hand, ovulation occurred spontaneously in rats after sympathetic denervation of the ovary by freezing its nerve supply (Wylie, Roche and Gibson 1985), so this route of stimulation is clearly not essential to release of the egg.

Once the oocyte within its cellular investment is exposed at the ovarian surface, it is propelled into the ostium of the tube by a coordinated activity of the cilia that densely line the fimbriated extremity of the ampulla. In many laboratory rodents and in horses there is an ovarian bursa, but in most of the domestic species used in studies of ovulation and egg transport (e.g. rabbit, sheep, pig), the

54

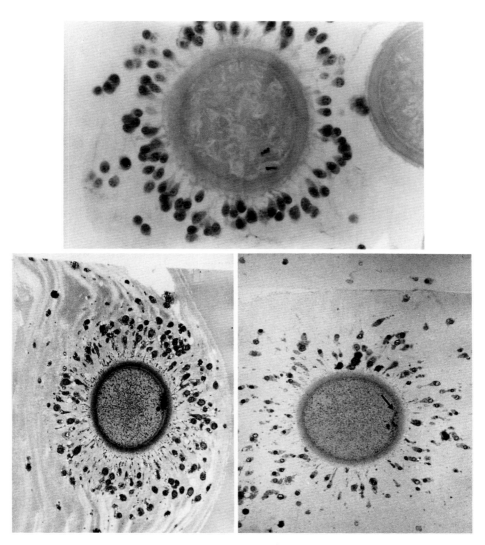

Fig. IV.1. Human oocytes invested with cells of the corona radiata showing the rearrangement of these cells and their cytoplasmic processes that occurs close to the time of ovulation. (Courtesy of Dr. Ondine Bomsel)

highly engorged fimbriated folds envelope and massage the ovaries at ovulation (see Fig. II.6). The latter process has been well illustrated by Blandau (1973), and is dependent upon the activity of specific fimbrial muscles and the supporting mesenteries. In this situation, the ciliated epithelium of the fimbriated infundibulum sweeps the follicular surface, and such local stimulation could be one of the explanations for the customary breakdown in the apical tissues of Graafian follicles at ovulation (Fléchon and Hunter 1980). The activity of these cilia is greatest at or just after the time of ovulation, and is known to be programmed

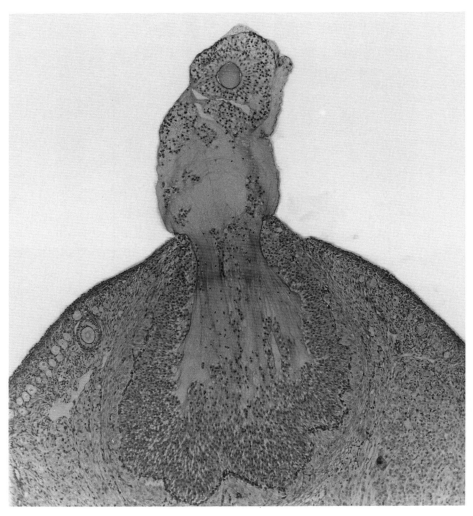

Fig. IV.2. Section through the ovulating follicle of a rabbit to show the liberated oocyte surrounded by its corona radiata being displaced in the escaping follicular fluid containing groups of granulosa cells. (Courtesy of Prof. R. J. Blandau)

by the prevailing levels of ovarian steroid hormones (Borell, Nilsson and Westman 1957; Blandau 1973). The cumulus cells have long been considered important in this initial phase of transport, enabling the tips of the cilia to obtain a purchase on the eggs, whereas denuded resin spheres used to simulate mammalian eggs simply rotate in situ on the fimbrial folds (Blandau 1969); cilia are unable to move them into the tubal ostium. Despite the inferred contrast between invested eggs and naked ones, cow and sheep eggs are thought to be almost denuded at ovulation or very soon thereafter (Lorton and First 1979), and the eggs of certain marsupials are liberated without any investment of follicular cells (Bedford 1983).

56

Once within the tubal ostium, eggs of many mammals pass to the site of fertilization at the ampullary-isthmic junction in a period of minutes. Estimates for cat and rabbit eggs, based in part on the study of vitally-stained cumulus plugs transferred on to the fimbrial folds, vary from 6–15 min (Harper 1961 a, b; Blandau 1969; Boling 1969), whereas measurements in the pig suggest an interval of 30–45 min (Andersen 1927; Hunter 1974). Irrespective of the precise timing in individual species, this initial phase of transport is remarkably rapid when compared with the total sojourn of the egg or embryo in the Fallopian tube; the latter is measured in days (see below). Moreover, the rapid phase of transport to the ampullary-isthmic junction indicates why fertilization usually occurs in that region rather than in more proximal regions of the tube. In the monkey, *Macaca nemestrina*, egg transport through the ampulla takes approximately 22 min (Blandau and Verdugo 1976), but 30 h or more are apparently required for the human egg to reach the ampullary-isthmic junction (Croxatto and Ortiz 1975; Pauerstein 1975; Eddy et al. 1976), indicating the ampulla as the site of fertilization. Despite the overall emphasis on rapidity in this phase of egg transport, there is clearly time for significant modifications in the egg membranes consequent upon the change of milieu from antral fluid in the follicle to secretions in the tubal lumen. Subtle alterations in oocyte metabolism might also be expected. After this initial transport, the eggs remain in the vicinity of the ampullary-isthmic junction for much of their tubal phase due to contraction or spasm of the musculature in this region.

Two principal mechanisms underlie egg transport to the site of fertilization: these are (1) the beat of cilia lining the ampullary epithelium and (2) waves of smooth muscle activity in the myosalpinx. Contractile activity in the neighbouring mesenteries may also be important (Blandau 1969, 1973), although it is not clear precisely how the influence of such contractions would be transmitted to the contents of the ampulla. Muscular and cilial activity in the Fallopian tubes responds to the circulating concentrations of ovarian steroid hormones, although of course the autonomic nervous system is directly regulating the myosalpinx. Once again, the rate of cilial beat and muscular contraction seems to be greatest just at or after the time of ovulation, thereby ensuring effective transport of eggs to the site of fertilization. However, progression of eggs to the ampullary-isthmic junction is not smooth and direct, but rather consists of intermittent rushes in step with waves of contraction in the ampulla (Harper 1961 a, b; Wintenberger-Torres 1961; Blandau 1973).

As discussed below with reference to the transport of spermatozoa, contractions and cilial beat in the isthmic portion of the tube are orientated towards the ovary at the time of ovulation. In one sense, therefore, the Fallopian tube can be viewed as two distinct compartments just after ovulation and around the time of fertilization: eggs are descending the ampulla whilst spermatozoa are ascending the isthmus. Because of the key rôle of ovarian steroids in regulating these events, anomalous levels of oestrogens and progesterone would certainly upset the normal mechanisms and timing of gamete transport.

A direct downwards displacement of the unfertilized egg by flow of fluid in the ampullary lumen is thought unlikely since the direction of bulk flow at the time of ovulation − albeit of a relatively small volume of fluid − is towards the peritoneal cavity rather than the uterus (Chap. III). In this situation, movement of eggs would be against the prevailing current, so a critical interaction between the cilial tips and the egg investments can be envisaged (Borell et al. 1957; Fléchon and Hunter 1981). If such contact did not exist, a far higher incidence of displacement of eggs by fluid flow into the peritoneal cavity would be expected. Even so, failure of appropriate cilial activity may underlie a proportion of ectopic pregnancies in women, although some patients with Kartagener's syndrome, the so-called 'immotile cilia syndrome', appear to be fertile (Afzelius, Camner and Mossberg 1978; Pauerstein and Eddy 1979). In this syndrome, the cilia are said to be completely immotile due to the lack of dynein arms. But when the absolute immotility of cilia in the genital — as distinct from respiratory (Kartagener 1933) — tract is confirmed, fertility may not be demonstrable. This would suggest that a functional ciliated endosalpinx is critical to human reproduction (McComb et al. 1986).

Deposition and Transport of Spermatozoa

The overall objective of sperm transport is to have a population of viable cells at or close to the site of fertilization shortly before the time of ovulation so that the egg (secondary oocyte) is penetrated before post-ovulatory ageing commences. Whilst this chapter will in due course focus on the distribution of spermatozoa within the Fallopian tubes, especially in relation to the time of ovulation, deposition and transport of spermatozoa in the lower reaches of the female tract must logically be considered first. Previous reviews on the topic of sperm transport in the female genital tract include those of Bedford (1970), Hunter (1973a, 1980, 1987a), Thibault (1973), Austin (1975), Overstreet and Katz (1977), Mortimer (1978, 1983), Polge (1978), Yanagimachi (1981), Harper (1982), Overstreet (1983) and Hawk (1987).

In perhaps a majority of eutherian mammals, semen is deposited in the anterior vagina at the time of mating; this is certainly the case in rabbits, ruminants (sheep, cows, goats, camels) and primates. However, in another group containing species as diverse as pig, horse, dog and various rodents, semen passes directly or indirectly into the uterus at the time of ejaculation. The precise situation depends on the anatomy of the penis and the arrangement of the external cervical tissues. In pigs, for example, the glans of the fibro-elastic penis has a spiral adaptation which enters the tapering oedematous folds of the cervix (Rigby 1967), and the principal stimulus for ejaculation is the interlocking so achieved. As a consequence of this arrangement and because of the large volume of semen, the uterine horns become distended with the ejaculate by the completion of a mating sequence — which may last for some 5–10 min or more (Hunter 1973a). In horses, a relatively large volume of semen also enters the uterus during mating but the tumescent penile shaft and glans penis do not intrude beyond the vagina. Rather, they serve to align the vagina and drooping (flaccid) cervix in a horizontal plane, enabling the contractions of ejaculation to propel the male fluids into the uterus (see Day 1942; Parker, Sullivan and First 1975).

Many of the small laboratory rodents such as rats, mice and hamsters have twin cervices, although these may fuse to give a single external os. The penis does not enter the cervical tissues, but semen is transmitted through the canals during rapid multiple copulations so that the uterine horns are distended with a mass of spermatozoa and fluid by the completion of mating (Hartman and Ball 1930; Blandau 1945). As would be expected, the nature of the sperm transport process in these species differs significantly from that found in rabbits, ruminants and primates. The last group will be considered first.

Intra-Vaginal Deposition. Upon ejaculation into the anterior vagina, the pool of highly concentrated semen characteristically bathes the external cervical os and spermatozoa gain the mucus-filled cervical canal largely under the influence of their own motility. Indeed, prior to associating with and penetrating the egg membranes, this is one of the two situations within the female duct system in which sperm motility is of paramount importance. (The other is when traversing the utero-tubal junction to enter the isthmus in species in which semen is deposited in the uterus.) This is not to argue that contractions of smooth muscle in the vaginal and cervical walls may not assist the process, but transport of spermatozoa en masse through the sometimes tortuous and mucus-filled cervical canal seems improbable. If it were a consistent feature, then displaced aggregates of mucus containing spermatozoa would have been a frequent observation in uterine flushings in experimental animals. Despite the predominant rôle of sperm motility in colonizing the cervix, the process is surprisingly rapid, at least for a vanguard of competent spermatozoa. This point is emphasized by the 93% incidence of fertilization obtained in rabbits after flushing the vagina with a detergent solution some 5 min from mating – a treatment that is highly effective in killing spermatozoa in the vagina itself (Bedford 1971). In a more recent series of experiments in rabbits, a species with a scant secretion of cervical mucus, Overstreet and his colleagues reported spermatozoa already in the Fallopian tubes within 1–2 min of mating (Overstreet and Cooper 1978; Overstreet, Cooper and Katz 1978), indicating that cervical transport must have been extremely rapid.

Interactions between spermatozoa and the cervical mucus have been examined in a number of species, especially in ruminants (Mattner 1963, 1966; Raynaud 1973) and in women (Moghissi 1977, 1984; Mortimer 1985). The physical condition of the mucus is critical in guiding a proportion of the spermatozoa to the cervical crypts, the point of origin of the mucus, since motile spermatozoa are orientated by the lines of strain in the anisotropic mucous secretion (Tampion and Gibbons 1962; Mattner 1966). Whether the cervical crypts in fact serve as a temporary reservoir for spermatozoa that subsequently escape and become actively involved in the events of fertilization is under a major question mark. Those spermatozoa that reach the Fallopian tubes to interact with the egg(s) are more probably ones that avoided the crypts by passing directly through the watery central channel in the cervical canal to enter the uterus (Fig. IV.3).

Spermatozoa tend to be highly concentrated within the vaginal pool of semen, certainly in comparison with cell density in species in which semen enters the uterus at mating (Table IV.2). One of the functions of the cervix is to reduce the density and initiate a gradient in absolute numbers of spermatozoa between the site of ejaculation and the site of fertilization. In simplified terms, the cervix imposes a reduction in density of at least 100-fold on the population of spermatozoa

CERVIX

VAGINA

Fig. IV.3. Diagrammatic representation of the cervix and its mucous secretion that originates from the crypts or glands. One current interpretation is that the spermatozoa gaining the Fallopian tubes to fertilize the eggs pass through the central, diluted channel of mucus directly to the uterus rather than becoming engaged in the cervical crypts

Table IV.2. Characteristics of ejaculated semen from mature, healthy animals to show the inverse relationship between the total volume of the ejaculate and the concentration of spermatozoa

Species	Volume of ejaculate (ml)	Expected sperm concentration ($\times 10^6$/ml)
Ram	0.8 – 1.2	2 000 – 3 000
Human	2 – 4	40 – 150
Bull	4 – 8	1 200 – 1 800
Stallion	30 – >150	100 – 250
Boar	150 – 500[a]	200 – 300

[a]Includes gelatinous secretion of bulbo-urethral glands.

entering the uterus (Fig. IV.4). Another function of the cervix is to enable spermatozoa to escape from the male secretions, the seminal plasma, so that those in the uterus are effectively resuspended in the fluids of this portion of the tract — a feature that may have important implications for sperm metabolism and for the final phases of maturation. Several experiments have demonstrated failure of seminal plasma to pass through the cervical canal into the uterus in species with intra-vaginal ejaculation (Walton 1930; Asch, Balmaceda and Pauerstein 1977), although this is not to infer the absence of an influence of seminal plasma constituents (e.g. prostaglandins) on the pattern of uterine contractions. But clinical observations in women undergoing artificial insemination indicate that whole

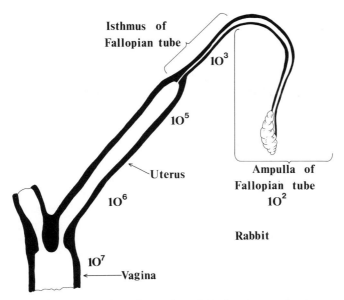

Isthmus of
Fallopian tube

10^3

10^5

Uterus

Ampulla of
Fallopian tube

10^2

10^6

Rabbit

10^7

Vagina

Fig. IV.4. Diagrammatic representation of a portion of the female reproductive tract of a rabbit to illustrate the steeply-declining gradient in sperm numbers found between the vagina and the site of fertilization in the Fallopian tube. (Redrawn after Austin 1965)

semen deposited directly into the body of the uterus usually leads to extremely painful spasm, which is clearly not the situation after coitus.

Actual movement of spermatozoa through the uterus to the Fallopian tubes depends on a combination of active and passive transport. Sperm motility (active transport) may be of direct value, although there is little persuasive evidence that such motility is direction-orientated towards the utero-tubal junction. Rather, the principal value of motility at this stage may be to maintain spermatozoa in suspension in the uterine fluids and prevent them adhering to epithelial surfaces (Austin 1964). Motility may also help to remove any residual elements of the male secretions not 'stripped' from the sperm plasma membrane during cervical passage; this is doubtless an important feature of the process of capacitation. Passive transport of spermatozoa within the uterus is due to the influence of myometrial contractions, which themselves may be heightened for a brief period under the influence of oxytocin secretion from the neurohypophysis; however, the half life of this hormone in the blood circulation is only 1−2 min. As noted above, smooth muscle stimulants in the seminal plasma such as prostaglandins may also contribute to contractile activity (Kelly et al. 1976; Poyser 1981). Some studies have claimed that myometrial contractions following mating early in oestrus are propagated largely towards the utero-tubal junction, as in sheep (Hawk 1975; Toutain et al. 1985). Nonetheless, in species with two distinct uterine horns such as cows, sheep and pigs, this claim must be reconciled with the observation that semen or marker dyes introduced into one horn shortly before ovulation can be redistributed to the contralateral horn (Rowson 1955; Polge 1978; Larsson 1986).

61

The timing of sperm transport through the cervix and uterus calls for specific comment even though, in a majority of mammals, mating will precede ovulation by a number of hours and therefore a requirement for rapid transport is not yet appreciated. Much of the established literature on sperm transport in the female tract falls into two schools of thought. On the one hand, spermatozoa are claimed to reach the site of fertilization in minutes (Van Demark and Hays 1954; Settlage, Motoshima and Tredway 1975, Overstreet and Katz 1977) while the opposing view is that the transport process requires a substantial number of hours (Dauzier 1958; Thibault 1973). But it is important to note that in none of these studies was the actual fertilizing potential of the transported spermatozoa assessed; judgements were simply based upon the observation of individual sperm cells, alive or dead, in smears, flushings or histological sections of the tubal contents or tissues. By contrast, when measurement is made of the length of time after mating for sufficient spermatozoa to have entered the Fallopian tubes subsequently to fertilize the eggs (i.e. competent spermatozoa), any putative phase of rapid transport appears not to be directly involved: slow, controlled transport is the demonstrably important component. Thus, some 6−8 h was the minimum period required for a population of functional spermatozoa to become established in the Fallopian tubes of sheep and cows mated at the onset of oestrus − as assessed by a surgical approach and subsequent examination of the eggs (Hunter, Nichol and Crabtree 1980; Hunter, Barwise and King 1982; Hunter and Wilmut 1983). The proportion of the 6−8 h required for traversing the cervix as distinct from the uterus is uncertain, but the figures nonetheless contrast starkly with those in minutes that are much-repeated in the literature for sperm transport to the Fallopian tubes, even though the process was accelerated in animals mated close to the time of ovulation. As to the extremely rapid phase of transport in rabbits referred to above, the spermatozoa so displaced were invariably moribund or dead (Overstreet and Cooper 1978), so the function of this process remains enigmatic. Apart from the possibility that it is some form of evolutionary vestige, two further suggestions are that: (1) dead or dying spermatozoa could release degradation products into the tubal lumen that interact with the epithelial secretions and the later-arriving spermatozoa and eggs to facilitate maturational changes in the gametes, or (2) they function to sensitize the peritoneal phagocytosis system in preparation for the arrival of greater numbers of motile spermatozoa (Hunter 1980; Hunter et al. 1980). In a similar vein, Overstreet (1983) has proposed that rapidly transported cells may be local messengers acting to coordinate movement of the reproductive tract.

Intra-Uterine Deposition. As already noted, semen is propelled into the uterus at mating in species such as pigs, horses, dogs and many rodents. Transport of spermatozoa as far as the utero-tubal junction should therefore not raise special problems, and a relatively rapid entry into the Fallopian tubes would be expected. Once more using the technique of estimating the time required for sufficient spermatozoa to enter the tubes in order subsequently to fertilize the eggs, values as low as 15 min have been reported for transport of a vanguard of spermatozoa in pigs (Hunter and Hall 1974a; Hunter 1981). The time required for adequate numbers of spermatozoa to enter the isthmus to ensure fertilization of all the eggs

– possibly 15–20 or more in this polytocous species – is usually less than one hour (Hunter 1981), but exceptionally may extend to 1–2 h (Hunter 1984). Similarly-derived functional estimates for sperm transport in mares and bitches do not yet exist, but long-standing values for the rat are 15–30 min between mating and the finding of spermatozoa in the ampullary portion of the tube (Blandau and Money 1944). Passage of adequate numbers of spermatozoa through the complex folds of the utero-tubal junction, a step largely dependent on cell motility (Gaddum-Rosse 1981), clearly requires a minimum threshold time.

Analogous to the rôle of the cervix in preventing access of seminal plasma to the uterus in species with intra-vaginal ejaculation, the utero-tubal junction acts to prevent gross passage of seminal plasma into the Fallopian tubes of species with intra-uterine ejaculation. This statement stems from chemical (Mann, Polge and Rowson 1956), physiological (Hunter and Hall 1974b) and radio-opaque tracer studies in pigs (Polge 1978). Even so, there is some suggestion from the use of radio-labelled inseminates that very small quantities of seminal plasma may in fact enter the caudal portion of the isthmus (Einarsson et al. 1980), possibly to influence the physiology of this portion of the tract.

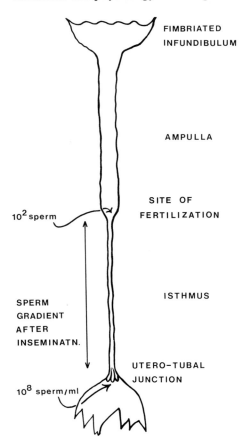

FIMBRIATED
INFUNDIBULUM

AMPULLA

SITE OF
FERTILIZATION

10^2 sperm

SPERM
GRADIENT
AFTER
INSEMINATN.

ISTHMUS

UTERO–TUBAL
JUNCTION

10^8 sperm/ml

Fig. IV.5. Representation of a Fallopian tube and tip of the uterine horn of a pig to portray the steeply declining sperm gradient from the utero-tubal junction to the site of fertilization

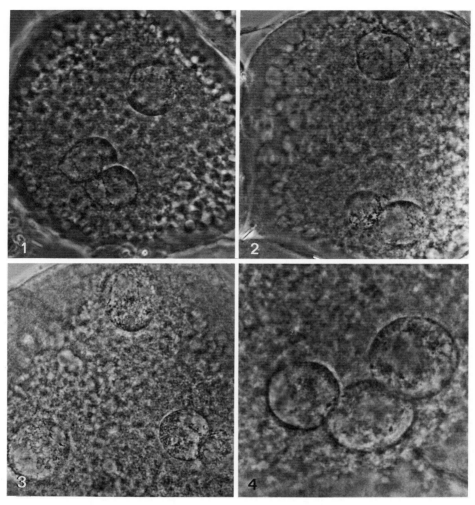

Fig. IV.6. Phase-contrast photomicrographs of whole-mount preparations of polyspermic pig eggs recovered from the Fallopian tube after experimental reduction of oedema in the mucosa and relaxation of the muscle layers in the isthmus. The female and a single male pronucleus are undergoing syngamy in 1 and 2, whilst an accessory male pronucleus is in the opposite hemisphere of the egg. A trispermic egg is shown in 3 whilst three pronuclei are at the onset of fusion in 4 (Hunter 1972)

Despite the lower cell concentration in the semen (Table IV.2), a major gradient in sperm numbers is essential in species with intra-uterine deposition; this gradient is imposed at the utero-tubal junction and in the isthmus (Fig. IV.5). The valve-like complexity of the former structure and the extremely narrow lumen of the isthmus together serve to reduce the numbers of spermatozoa potentially capable of achieving the site of fertilization. If the sperm gradient is overridden by means of experimental procedures, abnormal fertilization due to multiple

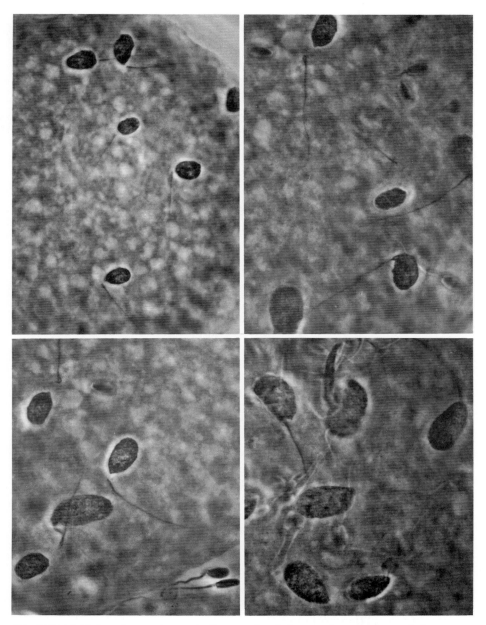

Fig. IV.7. Multiple polyspermic penetration seen in whole-mount preparations of pig eggs examined by phase-contrast microscopy. The polyspermic condition was induced by pre-ovulatory surgical insemination of sperm suspensions directly into the Fallopian tubes. Note that the sperm heads have separated from their respective mid-pieces and that there has been some degree of swelling of the sperm heads (Hunter 1976)

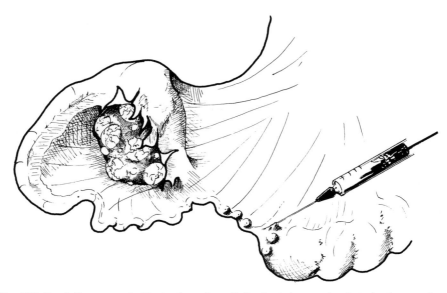

Fig. IV.8. Semi-diagrammatic illustration of one Fallopian tube, proximal uterine horn and portion of a pig ovary bearing pre-ovulatory follicles. Microdroplets of a solution of progesterone in oil injected beneath the serosal layer surrounding the distal isthmus and utero-tubal junction resulted in a 33% incidence of polyspermy in the eggs subsequently recovered from this tube (Hunter 1972)

penetration of the vitellus (polyspermy) is the usual sequel (Figs. IV.6 and IV.7). Such polyspermic penetration has been demonstrated in pigs by (1) introducing sperm suspensions directly into the Fallopian tubes at laparotomy (Polge, Salamon and Wilmut 1970; Hunter 1973b), (2) increasing the patency of the isthmus with local microinjections of progesterone to reduce oedema in the mucosa and thereby facilitate sperm passage (Hunter 1972; Fig. IV.8), and (3) by surgically resecting most of the isthmus and anastomosing the remaining portions of the tube (Hunter and Léglise 1971). Together, these experiments illustrate the necessity of a sperm gradient if the abnormal condition of polyspermic fertilization is to be avoided. Polyspermy is pathological in mammals and invariably leads to early death of the egg or embryo (Beatty 1957, 1961; Piko 1961; Austin 1963, 1965).

Pre- and Peri-Ovulatory Sperm Distribution Within the Isthmus

In terms of general transport mechanisms for spermatozoa already in the isthmus, peri-ovulatory waves of peristaltic contraction proceed principally towards the ampullary-isthmic junction (Blandau and Gaddum-Rosse 1974), and this activity should therefore assist transport of sperm cells. Such waves of contraction are propagated largely through the circular muscle, and may be sufficiently powerful to occlude the lumen and, for example, to transport droplets of oil or India ink to the peritoneal cavity, but the contractions tend to fade beyond the ampullary-isthmic junction (Blandau and Gaddum-Rosse 1974; Battalia and Yanagimachi 1979). As a quite distinct mechanism, Blandau and Gaddum-Rosse (1974) showed that cilia in the isthmus of rabbit and pig preparations beat actively towards the ampullary-isthmic junction at the time of ovulation, thereby creating local currents of fluid. Using an elegant in vitro system in which portions of the isthmus taken just before ovulation were sectioned longitudinally, these authors showed that *Lycopodium* spores or microspheres of approximately 15 μm diameter were transported actively by the cilial beat in a current of fluid, and they inferred that a similar influence may be exerted on spermatozoa. It should be pointed out, however, that such ciliary currents could not be detected in the isthmus of cat, cow, sheep, monkey or woman (Gaddum-Rosse, Blandau and Thiersch 1973).

Contractile activity in the isthmus may have a major influence on the distribution of spermatozoa relative to the time of ovulation. In spontaneously ovulating species such as sheep, cows and pigs mated at the onset of oestrus, and also in induced ovulators such as rabbits, there is persuasive evidence that viable spermatozoa are largely arrested in the distal portion of the Fallopian tubes until shortly before ovulation. A comparable restriction of spermatozoa to the isthmus has been described for mice (Zamboni 1972; Suarez 1987), rats (Shalgi and Kraicer 1978) and guinea-pigs (Yanagimachi and Mahi 1976). The inference here is that there may be a synchrony between release of the egg from the Graafian follicle and a redistribution of spermatozoa from the constricted lumen of the isthmus. In fact, for reasons developed below, the isthmus is now considered to act as the functional sperm reservoir, that is the reservoir from which spermatozoa are released at the time of ovulation to fertilize the eggs (Hunter 1986, 1987a; Hunter, Fléchon and Fléchon 1987). In addition to control of sperm progression by the patency of the pre-ovulatory isthmus, a significantly reduced beat of the sperm flagellum may be found in this storage site. Such a condition has been reported for rabbit and sheep spermatozoa (Overstreet and Cooper 1975; Overstreet and Katz 1977; Cummins 1982), and probably exists in other mammals. The composition of the pre-ovulatory luminal fluids seems to regulate sperm activity, for Burkman, Overstreet and Katz (1984) provided circumstantial evidence for differences in the composition of isthmic versus ampullary fluids in rabbits before ovulation: K^+ ions were thought to inhibit and pyruvate to stimulate sperm motility in the isthmus. Microenvironments within the Fallopian tubes may be demonstrated in other respects, for regional differences in pre-ovulatory temperatures have been recorded in rabbits (David, Vilensky and Nathan 1972) and pigs (Hunter and Nichol 1986), the distal isthmus being significantly cooler than the ampulla.

Chronological evidence for a pre-ovulatory regulation of sperm transport by the isthmus is as follows. By means of egg transplantation into mated or inseminated rabbit does that were themselves prevented from ovulating, Harper (1973a, b) demonstrated that viable spermatozoa were sequestered in the lower part of the tube for most of the pre-ovulatory interval, spermatozoa being released to fertilize the transplanted eggs only when ovulation in the recipient was imminent or had occurred. Harper suggested that the 'products of ovulation' − the eggs, their investments and/or the follicular fluid − entering the ampulla lead to movement of spermatozoa towards the site of fertilization. However, detailed

Table IV.3. The proportion of sheep yielding fertilized eggs and the incidence of fertilization when the interval from mating to transection of the isthmus 1.5 − 2.0 cm above the utero-tubal junction increased from 10 to 26 h[a]. (Taken from Hunter and Nichol 1983)

Interval from mating to transection (h)	Condition of ovaries at transection	No. of ewes		No. of eggs		No. of accessory sperm per egg	
		Examined	With some fertilized eggs	Recovered	Fertilized	Mean	Range
10	Pre-ovulatory	8	0	9	0	0	−
12	Pre-ovulatory	8	0	8	0	0	−
14	Pre-ovulatory	8	0	8	0	0	−
18	Pre-ovulatory	8	0	8	0	0	−
20	Pre-ovulatory	8	0	11	0	0	−
21	Pre-ovulatory	8	0	8	0	0	−
22	Pre-ovulatory	8	1	8	1	0	−
23	Pre-ovulatory	10	1	11	1	0	−
24	Pre-ovulatory	11	0	14	0	0	−
25	Peri-ovulatory	12	3	14	3	0.7	0 − 2
26	Post-ovulatory	13	11	16	13	7.9	0 − 26
Total		102	16	115	18	7.4	0 − 26

[a] All ewes operated on 26 h after the onset of oestrus had recent ovulations.

studies in farm animals with a pre-ovulatory interval longer than the 10 h found in rabbits show that spermatozoa commence to be displaced towards the site of fertilization only just before ovulation (Tables IV.3 and IV. 4). Thus, in ewes mated at the onset of oestrus, spermatozoa are arrested in the distal 1 − 2 cm of the isthmus for 17 − 18 h or more before peri-ovulatory release (Hunter and Nichol 1983), and in pigs similarly mated at the onset of oestrus, viable spermatozoa are very largely restricted to the distal 1 − 2 cm of the isthmus for up to 36 h or more of the 40-h pre-ovulatory interval (Hunter 1984). It would be instructive to know if a comparable regulation is found in bitches and mares with their pre-ovulatory oestrous periods of 4 − 7 days, and invaluable to establish if a similar control of sperm progression is found in primates. If the distal portion of the isthmus, in effect the intra-mural portion, does indeed act as a functional sperm reservoir in women, then this is relevant to procedures of reconstructive tubal surgery (Chap. VIII) and could bear on the practice of (1) artificial insemination and (2) the recovery and examination of tubal sperm populations in infertility clinics (Hunter 1987 b).

Programming of sperm release from the distal portion of the isthmus and their hyperactivation (Chap. V) may be linked to the maturity of the largest Graafian follicle(s), as expressed in the pattern of hormone secretion, although there is undoubtedly an interaction between ovarian hormones and the autonomic nervous supply to the myosalpinx. The objective would be to permit movement and displacement of regulated numbers of competent spermatozoa towards the site of fertilization closely synchronous with entry of egg(s) into the tubal ampulla. Because of the short fertilizable life of the mammalian egg, ovarian mechanisms that coordinate and stimulate release of spermatozoa from the lower regions of

Table IV.4. The influence of transecting the isthmus of pigs 1.5 – 2.0 cm proximal to the utero-tubal junction at increasing intervals after mating at the onset of oestrus on the proportion of eggs subsequently fertilized (6 animals/group). (Modified from Hunter 1984)

Interval from mating to transection (h)	Condition of ovaries at transection	Transected isthmus No. of		Control isthmus No. of	
		Eggs recovered	Eggs fertilized	Eggs recovered	Eggs fertilized
3	Pre-ovulatory	34	0	32	32
6	Pre-ovulatory	40	0	35	33
12	Pre-ovulatory	42	0	41	41
24	Pre-ovulatory	50	0	41	39
30	Pre-ovulatory	53	0	33	32
36	Pre-ovulatory	51	1	41	41
38	Pre-ovulatory	39	2	49	49
40	Peri-ovulatory	48	19	35	35
42 – 44	Post-ovulatory	46	46	34	34
Total		403	68	341	336

the tube would have obvious biological advantages, and there appear to be certain analogies with sperm storage and release in the avian oviduct. The actual means of endocrine control that has been proposed and demonstrated in pigs involves a local counter-current transfer of follicular hormones from the ovarian vein to the ovarian and utero-tubal arteries (Hunter, Cook and Poyser 1983). Elevated concentrations of steroids and prostaglandins transferred in this way would permit incisive changes in the patency and contractile activity of the isthmus, thereby facilitating the peri-ovulatory phase of sperm transport to the site of fertilization. Comparable vascular and/or lymphatic mechanisms may be involved in these peri-ovulatory events in sheep (see Walsh, Yutrzenka and Davis 1979) and women (Einer-Jensen, Bendz and Leidenberger 1985).

A further influence on sperm transport from the isthmus may occur at ovulation when follicular fluid with high concentrations of steroid hormones, several species of prostaglandin, and the peptide hormone relaxin is released into the tubal ampulla. This is not to infer that the follicular fluid passes down the Fallopian tubes but, because each ovary is tightly encompassed by the fimbriated infundibulum at ovulation, an initial contact of fluid with the ampulla would seem difficult to avoid.

Whilst the preceding paragraphs describe a means of programming the functional sperm reservoir in the isthmus, they do not explain the evolution of such a reservoir − except in terms of proximity to the ovarian control mechanism. Spermatozoa achieving this region of the duct system have undergone a form of selection for fitness (see below) and would be protected from phagocytosis, at least during the pre-ovulatory interval, by the population of polymorphs in the uterine lumen. Irrespective of the site of semen deposition at mating, spermatozoa in the isthmus would have escaped from the seminal plasma and may, as a consequence, be especially sensitive to storage mechanisms. Arrest of spermatozoa in a pre-capacitated state with intact plasma and acrosomal membranes would clearly be beneficial for the phase of storage and preservation of subsequent fertilizing ability (Hunter and Nichol 1983). It would seem logical, moreover, to locate the functional sperm reservoir within a short distance of the site of fertilization, so that activated spermatozoa can rapidly make contact with the egg membranes. Finally, this storage site may be a vestige of the situation in ancestral vertebrates in which chemotactic mechanisms were an essential feature of the fertilization process, before sophisticated ovarian endocrine mechanisms had developed.

Intra-Peritoneal Insemination

Whilst spermatozoa usually approach the Fallopian tubes from the lumen of the uterus, they may also gain the site of fertilization from the peritoneal cavity. This can occur spontaneously as a sequel to mating when spermatozoa leave the fimbriated extremity of one tube and enter the corresponding extremity of the other. Access from the peritoneal cavity is demonstrated experimentally by obstructing and/or transecting one uterine horn before mating virgin animals and subsequently recovering fertilized eggs from above the region of the obstruction, indicating a transperitoneal passage of viable spermatozoa from the patent tube. A more direct approach is to perform an intra-peritoneal insemination (in the absence of any previous mating), and subsequently to recover fertilized eggs. This intra-peritoneal route of insemination is feasible in cows (Skjerven 1955), guinea-pigs (Rowlands 1957), rabbits (Dauzier and Thibault 1956; Hadek 1958; Rowlands 1958) and pigs (Hunter 1978), and demonstrates that spermatozoa are able to retain their viability whilst descending the Fallopian tubes. Such a distribution of spermatozoa is achieved predominantly under the influence of cilial and muscular activity, but may be assisted by contact with the egg investments released at ovulation.

The experimental approach of intra-peritoneal insemination has clarified three aspects of tubal function. First, as judged from a detailed study in virgin pigs (gilts), the incidence of fertilization only approaches the expected post-coital levels (i.e. >90%: Hancock 1961) if the sperm suspension is introduced into the abdomen shortly before ovulation. Even then, the mean proportion of eggs fertilized was closer to 70% (Hunter 1978). The inference is that, in the absence of a sperm reservoir in the distal portion of the isthmus, timing of insemination is critical due to the short survival of spermatozoa in the tubal ampulla and/or peritoneal cavity. An alternative interpretation might be that it is the vigorous peri-ovulatory movement of the fimbria and its richly ciliated epithelium that is required to displace spermatozoa into the tubal ostium: spermatozoa presented to the fimbrial surface too late after the phase of intense peri-ovulatory activity may have a poor chance of entering the tube. Second, judgements on the time required for capacitation of spermatozoa (Chap. V) after intra-peritoneal insemination indicate no specific contribution of the peritoneal environment to this process of sperm maturation: the timing of capacitation was not reduced when com-

70

pared with that found after insemination directly into the Fallopian tubes. By contrast, the uterus and tubes act synergistically after mating to promote a more rapid achievement of capacitation (rabbit: Adams and Chang 1962; Bedford 1968; pig: Hunter and Hall 1974b). Third, boar spermatozoa gaining the ampulla after intra-peritoneal insemination and not becoming associated with the egg investments were not phagocytosed in the tubal lumen, but rather descended through the isthmus and entered the uterus with the eggs, presumably as a result of myosalpingeal activity (Hunter 1978). These results from intra-peritoneal insemination therefore illustrate the retrograde displacement of non-viable sperm cells into the uterus at the time of embryo descent. A comparable situation may be found after natural mating.

Sperm Selection in the Female Tract

The notion exists in much of the older literature that the population of spermatozoa reaching the Fallopian tubes may in some manner be selected during passage through the lower regions of the female tract (Parkes 1960; Austin 1964). In the first instance, the idea of selection probably stemmed from observations on the sperm gradient: compared with the numbers of spermatozoa present at the site of ejaculation, very few would be found in the ampullary region of the tubes during a single observation − hence the possibility of selection. In fact, studies using mixtures of differentially labelled live and dead spermatozoa (Baker and Degen 1972), frozen-thawed spermatozoa (Polge et al. 1970) or diploid marker spermatozoa (Mortimer 1977) have all strongly suggested that there is a progressive selection of cells between the site of deposition and the site of fertilization, but this could be interpreted simply as an elimination of less viable, moribund or dead spermatozoa, or those of inappropriate dimensions. Certainly, a higher proportion of morphologically normal spermatozoa is found in the isthmus than in the lower reaches of the tract.

Only the work of Cohen's group argues that there is some form of active selection operating in the female tract against spermatozoa with an imperfect genome, especially in the region of the utero-tubal junction. If the inference is that selection occurs on the basis of sperm surface characteristics, then this raises again the controversial question of haploid gene expression, for which there is now some positive evidence, at least for testicular spermatozoa (Hecht 1986). Cohen argues for selection of a small population of viable spermatozoa on the basis of calculations of chiasma frequency and sperm redundancy values in a range of mammals (Cohen 1967), and also from the results of studies with labelled spermatozoa in mice and rabbits (Cohen and McNaughton 1974; Cohen and Werrett 1975). The mechanisms whereby (a) an unrelated virgin female is programmed to select spermatozoa and (b) the specific detection and elimination of defective spermatozoa take place have not been clarified. An alternative explanation of Cohen's data might simply be that the populations of spermatozoa recovered from successively higher regions of the female tract were indeed 'fitter' in that they contained fewer, non-viable, poorly motile or moribund spermatozoa than those from the lower regions of the tract or ejaculate. As such they would, of course, be selected populations, but not ones produced by any active selection against genetically abnormal gametes by the female tract (Mortimer 1978, 1983).

Despite the balance of the preceding remarks, and the fact that more recent work from Cohen's group has not been in line with the original interpretations (Cohen and Tyler 1980), there remains evidence in mice that spermatozoa carrying certain t alleles have difficulty in traversing the utero-tubal junction (Braden and Gluecksohn-Waelsch 1958). But it is important to stress that some abnormal spermatozoa do reach the site of fertilization and may penetrate the eggs, albeit at a lower incidence (Nestor and Handel 1984). It is also worth noting that mouse spermatozoa carrying the t^{w32} complex have an accelerated rate of arrival in the ampulla (Tessler and Olds-Clarke 1981).

The presence in the lumen of the female genital tract of antibodies against spermatozoa or seminal plasma proteins may produce agglutination of large numbers of spermatozoa, thereby rendering them non-functional. However, this pathological condition cannot be viewed as a form of sperm selection in the sense of the preceding discussion.

Finally, once the site of fertilization is actually attained, spermatozoa still require considerable vigour to penetrate between the cells of any cumulus investment around the eggs. Acrosomal enzymes are also required for lysing a way through the egg membranes. Sperm viability and integrity therefore impose a rigorous form of selection at the egg surface.

Gamete Lifespans in the Female Tract

Although the lifespan of eggs and spermatozoa in the female reproductive tract is discussed in many reviews and monographs (e.g. Austin 1961, 1970; Thibault 1967; Szöllösi 1975; Gwatkin 1977; Yanagimachi 1981), a number of relevant questions have yet to receive satisfactory answers. First, however, there is a necessary clarification. The lifespan of gametes in the female tract should, strictly speaking, be viewed as the period of time during which the ability to undergo normal fertilization and give rise to a viable embryo is retained. In the case of the egg, this differs from the significantly greater timespan during which sperm penetration into the vitellus may occur, because ageing and indeed visibly degenerating eggs may still be capable of penetration (Blandau and Jordan 1941; Chang 1952; Braden 1959; Hunter 1967). This distinction between the functional lifespan of an egg and the 'penetrable' lifespan is not always clear in the literature. Moreover, in the case of spermatozoa, a population of millions of cells rather than a single gamete is invariably under consideration. This complicates the issue, for the ejaculate is known to contain a heterogeneous population of cells, some being dead or moribund and others immature with a retained cytoplasmic droplet (Mann 1964). Such heterogeneity may stem from various sources, including the dynamic nature of the process of spermatogenesis, the extent of epididymal storage, and the frequency of usage of the male, and itself must underlie the greater lifespan of the ejaculate.

With these points in mind, a specific question to raise is whether the viable lifespan of a gamete in the female tract is a rather constant figure for a given

Table IV.5. The incidence of polyspermic fertilization exhibited in mature pigs in various experimental situations after mating or insemination at the time of oestrus (Hunter 1979)

Treatment	Eggs examined No.	Polyspermic eggs	
		No.	%
Delayed mating	53	6	11.0[a]
Delayed mating	41	12	29.2
Delayed insemination	149	23	15.4
Tubal reconstructive surgery	34	11	32.4
Progesterone microinjections	198	64	32.3
Tubal insemination	77	26	33.8

[a] A further 21% of the eggs were considered digynic.

Table IV.6. The viable lifespan of gametes in the female reproductive tract, representing the period during which they are able to achieve normal fertilization and subsequent cleavage. Even so, a proportion of the eggs penetrated at the upper end of their lifespan may succumb to early embryonic loss

Species	Eggs (h)	Spermatozoa[a] (h)
Cow	10 − 12	24 − 48
Sheep	10 − 15	24 − 48
Pig	8 − 12	24 − 42
Horse	8 − 10	140
Human	8 − 10	24 − 72

[a] Motility is invariably retained for a much longer period than fertilizing ability except, perhaps, in the case of stallion spermatozoa.

species (as inferred by much of the literature), expressing an intrinsic property of the cell, or whether it might be influenced by the condition of the tract, especially the composition of the luminal fluids. The fluid environment is known to depend on changes in the secretion of ovarian steroid hormones, and just before ovulation there is a change from predominantly oestrogen synthesis by the Graafian follicle(s) to secretion of progesterone; this latter secretion increases with development of the corpora lutea. In species having a prolonged period of oestrus, the lifespan of spermatozoa in the pre-ovulatory situation is striking, with reported figures of 3 − 6 days in dogs (Griffiths and Amoroso 1939; Doak, Hall and Dale 1967) and even 5 − 6 days in mares (Day 1942; Burkhardt 1949). One interpretation here is that spermatozoa may have extended viability in the pre-ovulatory, oestrogen dominated tract. The same comment may be true for sperm survival in women, for there are suggested lifespans of 3 − 4 days (Ahlgren 1975; Edwards 1980; Gould, Overstreet and Hanson 1984). By contrast, survival of spermatozoa in a progesterone-dominated post-ovulatory environment would seem an unnecessary requirement. Even so, the diversity of nature is such that spermatozoa can apparently remain viable in the genital tract of hares (*Lepus europus*) for the duration of pregnancy (Martinet and Raynaud 1975) and perhaps also during pregnancy in certain strains of mice (Ullmann 1976).

A second question is whether all eggs ovulated spontaneously in the biological situation, as distinct from after a régime of steroid hormone therapy and/or gonadotrophin injection, are potentially viable and have similar lifespans in the Fallopian tubes. The usually-accepted idea is that there is a stringent selection of the follicle or follicles destined to ovulate and therefore the oocyte(s) so released is assumed to be viable. This would seem of critical importance to reproductive success in a monotocous species, and because of the dynamic interrelationships between oocyte and follicle (Foote and Thibault 1969; Thibault 1977), it may well be so. On the other hand, the incidence of spontaneous embryonic loss in many populations of mammals ($\geqslant 30 - 40\%$) and in women $40 - 60\%$ (Boué, Boué and Lazar 1975; Short 1979; Edwards 1980; Edmonds et al. 1982) raises the possibility that some ovulated oocytes may be defective. Relevant to this point is the fact that the incidence of polyspermic fertilization in pigs under diverse experimental treatments is $30 - 35\%$ (Table IV.5), which is remarkably similar to the proportion of prenatal loss. Accordingly, it is possible that the susceptibility to polyspermy in specified experimental conditions may be a means of revealing eggs of marginal viability that are destined to undergo early embryonic death (Hunter 1979).

With these various limitations in mind, values for gamete lifespans in the female genital tract are presented in Table IV.6. As will be noted, the egg commences to degenerate after only a small number of hours. The forms of degeneration are diverse, affecting many of the organelles in this remarkably large sphere of cytoplasm (Blandau 1975), but changes in (1) the microtubules of the meiotic spindle and (2) the location and integrity of the cortical granules may be especially significant.

REFERENCES

Adams CE, Chang MC (1962) Capacitation of rabbit spermatozoa in the Fallopian tube and in the uterus. J Exp Zool 15:159−166
Afzelius BA, Camner P, Mossberg B (1978) On the function of cilia in the female reproductive tract. Fertil Steril 29:72−74
Ahlgren M (1975) Sperm transport to and survival in the human Fallopian tube. Gynecol Invest 6:206−214
Aitken RJ, Kelly RW (1985) Analysis of the direct effects of prostaglandins on human sperm function. J Reprod Fertil 73:139−146
Amsterdam A, Lindner HR, Gröschel-Stewart U (1977) Localization of actin and myosin in the rat oocyte and follicular wall by immunofluorescence. Anat Rec 187:311−317
Andersen DH (1927) The rate of passage of the mammalian ovum through various portions of the Fallopian tube. Am J Physiol 82:557−569
Asch RH, Balmaceda J, Pauerstein CJ (1977) Failure of seminal plasma to enter the uterus and oviducts of the rabbit following artificial insemination. Fertil Steril 28:671−673
Austin CR (1961) The mammalian egg. Blackwell Sci Publ, Oxford
Austin CR (1963) Fertilization and transport of the ovum. In: Mechanisms concerned with conception. Pergamon Press, Oxford, pp 285−320
Austin CR (1964) Behaviour of spermatozoa in the female genital tract and in fertilization. Proc 5th Int Congr Anim Reprod, Trento 3:7−22
Austin CR (1965) Fertilization. Prentice-Hall, New Jersey
Austin CR (1970) Ageing and reproduction: post-ovulatory deterioration of the egg. J Reprod Fertil Suppl 12:39−53

Austin CR (1975) Sperm fertility, viability and persistence in the female tract. J Reprod Fertil Suppl 22:75–89

Baker RD, Degen AA (1972) Transport of live and dead boar spermatozoa within the reproductive tract of gilts. J Reprod Fertil 28:369–377

Battalia DE, Yanagimachi R (1979) Enhanced and coordinated movement of the hamster oviduct during the periovulatory period. J Reprod Fertil 56:515–520

Bauminger S, Lindner HR (1975) Periovulatory changes in ovarian prostaglandin formation and their hormonal control in the rat. Prostaglandins 9:737–751

Beatty RA (1957) Parthenogenesis and polyploidy in mammalian development. Cambridge University Press

Beatty RA (1961) Genetics of mammalian gametes. Anim Breed Abstr 29:243–256

Bedford JM (1968) Importance of the Fallopian tube for capacitation in the rabbit. Proc 6th Int Congr Anim Reprod, Paris 1:35–37

Bedford JM (1970) The saga of mammalian sperm from ejaculation to syngamy. Proc 21st Mosbach Symposium. Springer, Berlin Heidelberg New York

Bedford JM (1971) The rate of sperm passage into the cervix after coitus in the rabbit. J Reprod Fertil 25:211–218

Bedford JM (1983) Significance of the need for sperm capacitation before fertilization in eutherian mammals. Biol Reprod 28:108–120

Blandau RJ (1945) On the factors involved in sperm transport through the cervix uteri of the albino rat. Am J Anat 77:253–272

Blandau RJ (1969) Gamete transport – comparative aspects. In: Hafez ESE, Blandau RJ (eds) The mammalian oviduct. University of Chicago Press, Chicago, pp 129–162

Blandau RJ (1973) Gamete transport in the female mammal. In: Greep RO, Astwood EB (eds) Handbook of physiology, Section 7, Endocrinology II. American Physiological Society, Washington, pp 153–163

Blandau RJ (1975) Aging gametes: their biology and pathology. Karger, Basel

Blandau RJ (1978) Mechanism of tubal transport – comparative aspects. In: Brosens I, Winston R (eds) Reversibility of female sterilization. Academic Press, London, pp 1–15

Blandau RJ, Gaddum-Rosse P (1974) Mechanism of sperm transport in pig oviducts. Fertil Steril 25:61–67

Blandau RJ, Jordan ES (1941) The effect of delayed fertilization on the development of the rat ovum. Am J Anat 68:275–287

Blandau RJ, Money WL (1944) Observations on the rate of transport of spermatozoa in the female genital tract of the rat. Anat Rec 90:255–260

Blandau RJ, Verdugo P (1976) An overview of gamete transport – comparative effects. In: Harper MJK, Pauerstein CJ (eds) Symposium on ovum transport and fertility regulation. Scriptor, Copenhagen, pp 138–146

Boling JL (1969) Endocrinology of oviductal musculature. In: Hafez ESE, Blandau RJ (eds) The mammalian oviduct. University of Chicago Press, Chicago, pp 163–181

Bomsel-Helmreich O (1985) Ultrasound and the preovulatory human follicle. In: Clarke JR (ed) Oxford reviews of reproductive biology, vol 7. Clarendon, Oxford, pp 1–72

Borell U, Nilsson O, Westman A (1957) Ciliary activity in the rabbit Fallopian tube during oestrus and after copulation. Acta Obstet Gynecol Scand 36:22–28

Boué J, Boué A, Lazar P (1975) The epidemiology of human spontaneous abortions with chromosomal anomalies. In: Blandau RJ (ed) Ageing gametes, their biology and pathology. Karger, Basel, pp 330–348

Braden AWH (1959) Are nongenetic defects of the gametes important in the etiology of prenatal mortality? Fertil Steril 10:285–298

Braden AWH, Gluecksohn-Waelsch S (1958) Further studies of the effect of the T locus in the house mouse on male fertility. J Exp Zool 138:431–452

Brenner RM (1973) Endocrine control of ciliogenesis in the primate oviduct. In: Greep RO, Astwood EB (eds) Handbook of physiology, Section 7, Endocrinology, vol 2. American Physiological Society, Washington, pp 123–139

Burkhardt J (1949) Sperm survival in the genital tract of the mare. J Agric Sci 39:201–203

Burkman LJ, Overstreet JW, Katz DF (1984) A possible role for potassium and pyruvate in the modulation of sperm motility in the rabbit oviducal isthmus. J Reprod Fertil 71:367–376

Chang MC (1952) Effects of delayed fertilization on segmenting ova, blastocysts and foetuses in rabbits. Fed Proc Am Socs Exp Biol 11:24

Cohen J (1967) Correlation between sperm 'redundancy' and chiasma frequency. Nature (Lond) 215:862–863

Cohen J, McNaughton DC (1974) Spermatozoa: the probable selection of a small population by the genital tract of the female rabbit. J Reprod Fertil 39:297–310

Cohen J, Tyler KR (1980) Sperm populations in the female genital tract of the rabbit. J Reprod Fertil 60:213–218

Cohen J, Werrett DJ (1975) Antibodies and sperm survival in the female tract of the mouse and rabbit. J Reprod Fertil 42:301–310

Croxatto HB, Ortiz MES (1975) Egg transport in the Fallopian tube. Gynecol Invest 6:215–225

Cummins JM (1982) Hyperactivated motility patterns of ram spermatozoa recovered from the oviducts of mated ewes. Gamete Res 6:53–63

Dauzier L (1958) Physiologie du déplacement des spermatozoides dans les voies génitales femelles chez la brebis et la vache. Annls Zootech 7:281–306

Dauzier L, Thibault C (1956) Recherche expérimentale sur la maturation des gamètes males chez les mammifères par l'étude de la fécondation in vitro de l'oeuf de lapine. Proc 3rd Int Congr Anim Reprod, Cambridge 1:58–61

David A, Vilensky A, Nathan H (1972) Temperature changes in the different parts of the rabbit's oviduct. Int J Gynecol Obstet 10:52–56

Day FT (1942) Survival of spermatozoa in the genital tract of the mare. J Agric Sci 32:108–111

Dekel N, Phillips DM (1979) Maturation of the rat cumulus oophorus: a scanning electron microscopic study. Biol Reprod 21:9–18

Doak RL, Hall A, Dale HE (1967) Longevity of spermatozoa in the reproductive tract of the bitch. J Reprod Fertil 13:51–58

Eddy CA, Garcia RG, Kraemer DC, Pauerstein CJ (1976) Ovum transport in non-human primates. In: Harper MJK, Pauerstein CJ (eds) Symposium on ovum transport and fertility regulation. Scriptor, Copenhagen, pp 390–403

Edmonds DK, Lindsay KS, Miller JF, Williamson E, Wood PJ (1982) Early embryonic mortality in women. Fertil Steril 38:447–453

Edwards RG (1980) Conception in the human female. Academic Press, London

Einarsson S, Jones B, Larsson K, Viring S (1980) Distribution of small- and medium-sized molecules within the genital tract of artificially inseminated gilts. J Reprod Fertil 59:453–457

Einer-Jensen N, Bendz A, Leidenberger F (1985) Transfer of 13-C progesterone in the human ovarian adnex. Proc Soc Stud Fertil Abstract 53:41

Eppig JJ (1980) Regulation of cumulus oophorus expansion by gonadotropin in vivo and in vitro. Biol Reprod 23:545–552

Eppig JJ (1982) The relationship between cumulus cell-oocyte coupling, oocyte meiotic maturation, and cumulus expansion. Develop Biol 89:268–272

Fléchon JE, Hunter RHF (1980) Ovulation in the pig: a light and scanning electron microscope study. Proc Soc Stud Fertil Abstract 5:9

Fléchon JE, Hunter RHF (1981) Distribution of spermatozoa in the utero-tubal junction and isthmus of pigs, and their relationship with the luminal epithelium after mating. Tissue & Cell 13:127–139

Foote WD, Thibault C (1969) Recherches expérimentales sur la maturation in vitro des ovocytes de truie et de veau. Annls Biol Anim Biochim Biophys 9:329–349

Gaddum-Rosse P (1981) Some observations on sperm transport through the uterotubal junction of the rat. Am J Anat 160:333–341

Gaddum-Rosse P, Blandau RJ, Thiersch JB (1973) Ciliary activity in the human and *Macaca nemestrina* oviduct. Am J Anat 138:269–275

Gould JE, Overstreet JW, Hanson FW (1984) Assessment of human sperm function after recovery from the female reproductive tract. Biol Reprod 31:888–894

Griffiths WFB, Amoroso EC (1939) Pro-oestrus, oestrus, ovulation and mating in the greyhound bitch. Vet Rec 51:1279–1284

Gwatkin RBL (1977) Fertilisation mechanisms in man and mammals. Plenum Press, New York

Hadek R (1958) Intraperitoneal insemination of rabbit doe. Proc Soc Exp Biol Med 99:39–40

Hancock JL (1961) Fertilization in the pig. J Reprod Fertil 2:307–331

Harper MJK (1961a) The mechanisms involved in the movement of newly ovulated eggs through the ampulla of the rabbit Fallopian tube. J Reprod Fertil 2:522–524

Harper MJK (1961b) Egg movement through the ampullar region of the Fallopian tube of the rabbit. Proc 4th Int Congr Anim Reprod, The Hague, p 375

Harper MJK (1973a) Stimulation of sperm movement from the isthmus to the site of fertilization in the rabbit oviduct. Biol Reprod 8:369–377

Harper MJK (1973b) Relationship between sperm transport and penetration of eggs in the rabbit oviduct. Biol Reprod 8:441–450

Harper MJK (1982) Sperm and egg transport. In: Austin CR, Short RV (eds) Reproduction in mammals, book 1. Cambridge University Press, pp 102–127

Hartman CG, Ball J (1930) On the almost instantaneous transport of spermatozoa through the cervix and uterus of the rat. Proc Soc Exp Biol Med 28:312–314

Hawk HW (1975) Hormonal control of changes in the direction of uterine contractions in the estrous ewe. Biol Reprod 12:423–430

Hawk HW (1987) Transport and fate of spermatozoa after insemination of cattle. J Dairy Sci 70:1487–1503

Hecht NB (1986) Haploid gene expression in the mammalian testis. Develop Growth Differentn 28, Suppl pp 33–34

Hunter RHF (1967) The effects of delayed insemination on fertilization and early cleavage in the pig. J Reprod Fertil 13:133–147

Hunter RHF (1972) Local action of progesterone leading to polyspermic fertilization in pigs. J Reprod Fertil 31:433–444

Hunter RHF (1973a) Transport, migration and survival of spermatozoa in the female genital tract: species with intra-uterine deposition of semen. In: Hafez ESE, Thibault C (eds) Sperm transport, survival and fertilizing ability. INSERM, Paris, pp 309–342

Hunter RHF (1973b) Polyspermic fertilization in pigs after tubal deposition of excessive numbers of spermatozoa. J Exp Zool 183:57–64

Hunter RHF (1974) Chronological and cytological details of fertilization and early embryonic development in the domestic pig, *Sus scrofa*. Anat Rec 178:169–186

Hunter RHF (1976) Sperm-egg interactions in the pig: monospermy, extensive polyspermy and the formation of chromatin aggregates. J Anat 122:43–59

Hunter RHF (1978) Intraperitoneal insemination, sperm transport and capacitation in the pig. Anim Reprod Sci 1:167–179

Hunter RHF (1979) Ovarian follicular responsiveness and oocyte quality after gonadotrophic stimulation of mature pigs. Annls Biol Anim Biochim Biophys 19:1511–1520

Hunter RHF (1980) Transport and storage of spermatozoa in the female tract. Proc 9th Int Congr Anim Reprod, Madrid 2:227–233

Hunter RHF (1981) Sperm transport and reservoirs in the pig oviduct in relation to the time of ovulation. J Reprod Fertil 63:109–117

Hunter RHF (1984) Pre-ovulatory arrest and peri-ovulatory redistribution of competent spermatozoa in the isthmus of the pig oviduct. J Reprod Fertil 72:203–211

Hunter RHF (1986) Peri-ovulatory physiology of the oviduct, with special reference to sperm transport, storage and capacitation. Develop Growth Differentn 28, Suppl pp 5–7

Hunter RHF (1987a) Peri-ovulatory physiology of the oviduct, with special reference to progression, storage and capacitation of spermatozoa. In: Mohri H (ed) New horizons in sperm cell research. Jpn Sci Soc, Tokyo, pp 31–45

Hunter RHF (1987b) Human fertilisation in vivo, with special reference to progression, storage and release of competent spermatozoa. Human Reprod 2, No 4:329–332

Hunter RHF, Hall JP (1974a) Capacitation of boar spermatozoa: the influence of post-coital separation of the uterus and Fallopian tubes. Anat Rec 180:597–604

Hunter RHF, Hall JP (1974b) Capacitation of boar spermatozoa: synergism between uterine and tubal environments. J Exp Zool 188:203–214

Hunter RHF, Léglise PC (1971) Polyspermic fertilization following tubal surgery in pigs, with particular reference to the rôle of the isthmus. J Reprod Fertil 24:233–246

Hunter RHF, Nichol R (1983) Transport of spermatozoa in the sheep oviduct: preovulatory sequestering of cells in the caudal isthmus. J Exp Zool 228:121–128

Hunter RHF, Nichol R (1986) A preovulatory temperature gradient between the isthmus and ampulla of pig oviducts during the phase of sperm storage. J Reprod Fertil 77:599–606

Hunter RHF, Wilmut I (1983) The rate of functional sperm transport into the oviducts of mated cows. Anim Reprod Sci 5:167–173

Hunter RHF, Nichol R, Crabtree SM (1980) Transport of spermatozoa in the ewe: timing of the establishment of a functional population in the oviduct. Reprod Nutr Develop 20:1869–1875

Hunter RHF, Barwise L, King R (1982) Sperm transport, storage and release in the sheep oviduct in relation to the time of ovulation. Brit Vet J 138:225–232

Hunter RHF, Cook B, Poyser NL (1983) Regulation of oviduct function in pigs by local transfer of ovarian steroids and prostaglandins: a mechanism to influence sperm transport. Europ J Obstet Gynecol Reprod Biol 14:225–232

Hunter RHF, Fléchon B, Fléchon JE (1987) Pre- and peri-ovulatory distribution of viable spermatozoa in the pig oviduct: a scanning electron microscope study. Tissue & Cell 19:423–436

Kartagener M (1933) Zur Pathogenese der Bronchiektasien. I Bronchiektasien bei Situs viscerum inversus. Beitr Klin Tuberk 83:489–501

Kelly RW, Taylor PL, Hearn JP, Short RV, Martin DE, Marston JH (1976) 19-Hydroxyprosta-glandin E_1 as a major component of the semen of primates. Nature (Lond) 260:544–545

Larsson B (1986) Transuterine transport of spermatozoa after artificial insemination in heifers. Anim Reprod Sci 12:115–122

LeMaire WJ, Yang NST, Behrman HR, Marsh JM (1973) Preovulatory changes in the concentration of prostaglandins in rabbit Graafian follicles. Prostaglandins 3:367–376

Lorton SP, First NL (1979) Hyaluronidase does not disperse the cumulus oophorus surrounding bovine ova. Biol Reprod 21:301–308

McComb P, Langley L, Villalon M, Verdugo P (1986) The oviductal cilia and Kartagener's syndrome. Fertil Steril 46:412–416

Mann T (1964) The biochemistry of semen and of the male reproductive tract. Methuen, London

Mann T, Polge C, Rowson LEA (1956) Participation of seminal plasma during the passage of spermatozoa in the female reproductive tract of the pig and horse. J Endocrinol 13:133–140

Martinet L, Raynaud F (1975) Prolonged spermatozoan survival in the female hare uterus: explanation of superfetation. In: Hafez ESE, Thibault CG (eds) The biology of spermatozoa. Karger, Basel, pp 134–144

Mattner PE (1963) Spermatozoa in the genital tract of the ewe. II Distribution after coitus. Austr J Biol Sci 16:688–694

Mattner PE (1966) Formation and retention of the spermatozoan reservoir in the cervix of the ruminant. Nature (Lond) 212:1479–1480

Moghissi KS (1977) Sperm migration through the human cervix. In: Insler V, Bettendorf G (eds) The uterine cervix in reproduction. Thieme, Stuttgart, p 146

Moghissi KS (1984) The function of the cervix in human reproduction. Curr Prob Obstet Gynecol 7:1

Mortimer D (1977) The survival and transport to the site of fertilization of diploid rabbit spermatozoa. J Reprod Fertil 51:99–104

Mortimer D (1978) Selectivity of sperm transport in the female genital tract. In: Cohen J, Hendry WF (eds) Spermatozoa, antibodies and infertility. Blackwells, Oxford, pp 37–53

Mortimer D (1983) Sperm transport in the human female reproductive tract. Oxford Rev Reprod Biol 5:30–61

Mortimer D (1985) The male factor in infertility. Part II, Sperm function testing. Curr Prob Obstet Gynecol 8:1–75

Nestor A, Handel MA (1984) The transport of morphologically abnormal sperm in the female reproductive tract of mice. Gamete Res 10:119–125

O'Shea JD, Phillips RE (1974) Contractility of ovarian follicles from sheep in vitro. J Reprod Fertil 36:457

Overstreet JW (1983) Transport of gametes in the reproductive tract of the female mammal. In: Hartmann JF (ed) Mechanism and control of animal fertilization. Academic Press, New York, pp 499–543

Overstreet JW, Cooper GW (1975) Reduced sperm motility in the isthmus of the rabbit oviduct. Nature (Lond) 258:718–719

Overstreet JW, Cooper GW (1978) Sperm transport in the reproductive tract of the female rabbit. I. The rapid transit phase of transport. Biol Reprod 19:101–114

Overstreet JW, Katz DF (1977) Sperm transport and selection in the female genital tract. In: Johnson MH (ed) Development in mammals, vol 2. Elsevier, Amsterdam, pp 31–65

Overstreet JW, Cooper GW, Katz DF (1978) Sperm transport in the reproductive tract of the female rabbit. II. The sustained phase of transport. Biol Reprod 19:115–132

Owman CH, Sjöberg N-O, Svensson K-G, Walles B (1975) Autonomic nerves mediating contractility in the human Graafian follicle. J Reprod Fertil 45:553–556

Parker WG, Sullivan JJ, First NL (1975) Sperm transport and distribution in the mare. J Reprod Fertil, Suppl 23:63–66

Parkes AS (1960) The biology of spermatozoa and artificial insemination. In: Parkes AS (ed) Marshall's physiology of reproduction. Longmans Green, London, pp 161–263

Pauerstein CJ (1975) Clinical implications of oviductal physiology and biochemistry. Gynecol Invest 6:253–264

Pauerstein CJ, Eddy CA (1979) The role of the oviduct in reproduction: our knowledge and our ignorance. J Reprod Fertil 55:223–229

Piko L (1961) La polyspermie chez les animaux. Annls Biol Anim Biochim Biophys 1:323–383

Polge C (1978) Fertilization in the pig and horse. J Reprod Fertil 54:461–470

Polge C, Salamon S, Wilmut I (1970) Fertilizing capacity of frozen boar semen following surgical insemination. Vet Rec 87:424–428

Poyser NL (1981) Prostaglandins in reproduction. Wiley & Sons, Chichester

Raynaud F (1973) Physiologie du col de l'utérus de la brebis. Effet d'un progestagène de synthèse: l'acétate de fluorogestone. Annls Biol Anim Biochim Biophys 13:335–346

Rigby JP (1967) The cervix of the sow during oestrus. Vet Rec 80:672–675

Rowlands IW (1957) Insemination of the guinea pig by intraperitoneal injection. J Endocrinol 16:98–106

Rowlands IW (1958) Insemination by intraperitoneal injection. Proc Soc Stud Fertil 10:150–157

Rowson LEA (1955) The movement of radio opaque material in the bovine uterine tract. Brit Vet J 111:334–342

Settlage DSF, Motoshima M, Tredway DR (1975) Sperm transport from the vagina to the Fallopian tubes in women. In: Hafez ESE, Thibault CG (eds) The biology of spermatozoa. Karger, Basel, pp 74–82

Shalgi R, Kraicer PF (1978) Timing of sperm transport, sperm penetration and cleavage in the rat. J Exp Zool 204:353–360

Short RV (1979) When a conception fails to become a pregnancy. In: Whelan J (ed) Maternal recognition of pregnancy. Excerpta Medica/North Holland, pp 377–387 (Ciba Fdn Colloq No 64)

Skjerven O (1955) Conception in a heifer after deposition of semen in the abdominal cavity. Fertil Steril 6:66–67

Suarez SS (1987) Sperm transport and motility in the mouse oviduct: observations in situ. Biol Reprod 36:203–210

Szöllösi D (1975) Mammalian eggs aging in the Fallopian tubes. In: Blandau RJ (ed) Aging gametes: their biology and pathology. Karger, Basel, pp 98–121

Szöllösi D, Gérard M (1983) Cytoplasmic changes in the mammalian oocytes during the preovulatory period. In: Beier HM, Lindner HR (eds) Fertilization of the human egg in vitro. Springer, Berlin Heidelberg New York, pp 35–55

Tampion D, Gibbons RA (1962) Orientation of spermatozoa in mucus of the cervix uteri. Nature (Lond) 194:381

Tessler S, Olds-Clarke P (1981) Male genotype influences sperm transport in female mice. Biol Reprod 24:806–813

Thibault C (1967) Analyse comparée de la fécondation chez la brebis, la vache et la lapine. Annls Biol Anim Biochim Biophys 7:5–23

Thibault C (1973) Sperm transport and storage in vertebrates. J Reprod Fertil Suppl 18:39–53

Thibault C (1977) Are follicular maturation and oocyte maturation independent processes? J Reprod Fertil 51:1‒15

Toutain PL, Marnet PG, Laurentie MP, Garcia-Villar R, Ruckebusch Y (1985) Direction of uterine contractions during estrus in ewes: a re-evaluation. Am J Physiol 249:R410‒R416

Ullman SL (1976) Anomalous litters in hybrid mice and the retention of spermatozoa in the female tract. J Reprod Fertil 47:13‒18

Van Demark NL, Hays RL (1954) Rapid sperm transport in the cow. Fertil Steril 5:131‒137

Walles B, Owman CH, Sjöberg N-O (1982) Contraction of the ovarian follicle induced by local stimulation of its sympathetic nerves. Brain Res Bull 9:757‒760

Walsh SW, Yutrzenka GJ, Davis JS (1979) Local steroid concentrating mechanism in the reproductive vasculature of the ewe. Biol Reprod 20:1167‒1171

Walton A (1930) On the function of the rabbit cervix during coitus. J Obstet Gynecol Brit Emp 37:92‒95

Wintenberger-Torres S (1961) Mouvements des trompes et progression des oeufs chez la brebis. Annls Biol Anim Biochim Biophys 1:121‒133

Wylie SN, Roche PJ, Gibson WR (1985) Ovulation after sympathetic denervation of the rat ovary produced by freezing its nerve supply. J Reprod Fertil 75:369‒373

Yanagimachi R (1981) Mechanisms of fertilization in mammals. In: Mastroianni L, Biggers JD (eds) Fertilization and embryonic development in vitro. Plenum, New York, pp 81‒182

Yanagimachi R, Mahi C (1976) The sperm acrosome reaction and fertilization in the guinea-pig: a study in vivo. J Reprod Fertil 46:49‒54

Zamboni L (1972) Fertilization in the mouse. In: Moghissi KS, Hafez ESE (eds) Biology of mammalian fertilization and implantation. Thomas, Springfield, Illinois, pp 213‒262

Denudation of Eggs, Capacitation of Spermatozoa, and Fertilization – Normal and Abnormal

CONTENTS

Introduction and Chemotaxis 81
Denudation and Final Maturation of Eggs 82
Fertilizin-Antifertilizin Reactions at the Egg Surface 84
Binding Sites for Spermatozoa on the Zona Pellucida 85
Capacitation of Spermatozoa, the Acrosome Reaction and Hyperactivation 88
Penetration of the Zona and Fusion of Gametes 96
Activation of the Egg and Block to Polyspermy 97
Synchrony of Penetration and Cleavage 99
Abnormalities of Fertilization 101
Fate of Redundant Gametes 102
References 103

Introduction and Chemotaxis

Whilst the previous chapter has discussed transport of gametes to the site of fertilization in the Fallopian tubes, actual penetration of the egg membranes requires final maturational changes in the sperm cell. There may also be maturational changes in the liberated secondary oocyte, quite apart from in its investment of follicular cells, but the evidence for such changes remains equivocal. One of the most intriguing topics related to these events is that of *chemotaxis* defined, for present purposes, as release of a specific substance by the egg surface to attract spermatozoa and thereby to promote the efficiency of fertilization. There is no convincing evidence for the existence of this phenomenon in eutherian mammals (Yanagimachi 1981), although lack of such evidence does not, of course, mean that chemotaxis does not exist. On the contrary, the experimental systems used so far for exposing spermatozoa in vitro to eggs or their surface fluids may have been inappropriate or simply insufficiently sensitive to reveal the phenomenon. A more plausible explanation, however, is that mechanisms other than chemotaxis have evolved within the duct systems of placental mammals to foster a timely and effective contact of eggs and spermatozoa leading to successful fertilization. The ducts are under the influence of ovarian hormones and a local means of control by the pre-ovulatory Graafian follicle(s) has already been proposed (Chap. IV). Further regulation by the autonomic nervous system and locally released catecholamines may serve to coordinate the meeting of eggs and spermatozoa.

Denudation refers to loss of follicular cells from the surface of the zona pellucida in the lumen of the Fallopian tube. In unmated animals, the process occurs spontaneously but it is thought to be accelerated in the presence of spermatozoa. The extent of denudation is clearly related to the condition of the egg(s) at ovulation. Those of the pig, rabbit, many laboratory rodents and women enter the tube invested with cumulus cells. In polyovular species, cumulus masses surrounding individual eggs usually aggregate to form a cumulus "plug" (Fig. V.1) for transport to the ampullary-isthmic junction (Hancock 1962). By contrast, eggs of the cow and to a lesser extent those of sheep, enter the tube almost completely denuded of follicular cells (Chap. IV).

In Graafian follicles responding to a pre-ovulatory surge of gonadotrophic hormones, rearrangement of the cells surrounding the oocyte commences some hours before ovulation. The cells immediately adjoining the zona pellucida, those of the corona radiata, show a distinct morphological change: their corona cell processes are gradually withdrawn from the substance of the zona and the cells

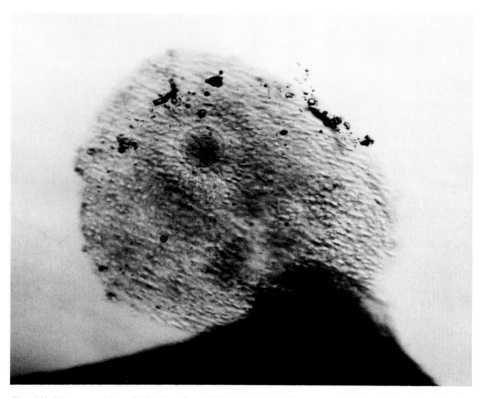

Fig. V.1. The apex of a rabbit Graafian follicle just after the moment of ovulation showing the newly released egg surrounded by granulosa cells. Eggs so arranged may aggregate to form a cumulus 'plug'. (Courtesy of Dr. M. J. K. Harper)

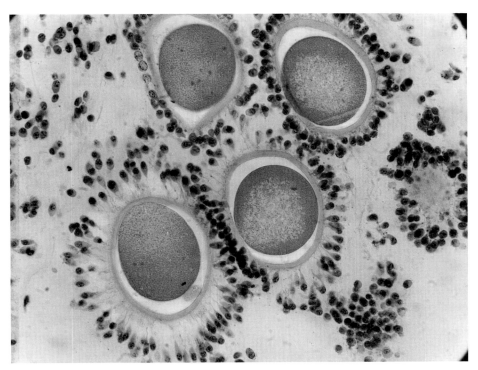

Fig. V.2. The morphological rearrangement of cells of the corona radiata around rabbit oocytes close to the time of ovulation. Note especially the elongated corona cell processes. (Courtesy of Prof. C. Thibault; after Hunter 1980)

themselves assume a radially-arranged appearance with elongated processes (Fig. V.2). The surrounding granulosa cells, generally referred to as those of the cumulus oophorus, also show a progressive loosening and expansion (Chap. IV). Indeed, expansion of the cumulus mass has been used in echo-sonography studies of human Graafian follicles as a criterion of selection for laparoscopic puncture and oocyte aspiration (Bomsel-Helmreich 1985; Demoulin 1985). Progressive dispersal of cumulus cells in the Fallopian tube may be regarded simply as a continuation of events commencing in the pre-ovulatory follicle, although at least one report has claimed that the bicarbonate ion in tubal fluid promotes the denudation process (Stambaugh, Noriega and Mastroianni 1969). Further subtle biochemical activities doubtless also contribute to denudation, but a purely mechanical influence of interactions with cilia must not be overlooked, nor the constant squeezing and deformation by contractile activity in the myosalpinx during egg descent to the site of fertilization.

The acceleration of oocyte denudation in the presence of spermatozoa referred to above concerns an action of sperm hyaluronidase on the hyaluronic acid cement substance that binds follicular cells. The enzyme is thought to be released physiologically as a result of membrane vesiculation around the acrosome (Austin 1963a), but dead or dying spermatozoa may also liberate their complement of proteolytic enzymes which could then act to depolymerize the hyaluronic acid and so disperse the follicular cells. At least one report proposes a release of hyaluronidase after an initial swelling of the acrosome but before the vesiculation reaction is detectable (Szollozi and Hunter 1978); this observation concerned boar spermatozoa, and it may therefore be relevant to emphasize the extensive cumulus plug around freshly-shed pig eggs (Hancock 1961).

In the context of fertilization, one suggestion of long standing is that the follicular cells surrounding the secondary oocyte serve to increase the surface area and thus the potential target for spermatozoa (Chang and Pincus 1951; Austin and Walton 1960; Austin 1961). This would be especially valuable in the absence of any chemotactic mechanism, for the follicular investment could be viewed as a means of trapping motile spermatozoa. If this line of reasoning is correct, the pattern of sperm transport within the Fallopian tubes may be related to the condition of the egg(s) at ovulation: more spermatozoa might be expected to ascend the isthmus in species in which the egg is essentially denuded at ovulation. This would enhance the probability of sperm-egg encounters whilst still avoiding too great a risk of polyspermy. However, detailed studies in sheep and cattle reveal rather similar peri-ovulatory patterns of sperm distribution in the isthmus when compared with those in pigs.

Discussions of denudation usually focus on the oocyte and neglect the fact that liberated follicular cells are not necessarily becoming redundant or being destroyed; they may have important functions in the tube after release from the egg surface. Since follicular cells can be successfully cultured in vitro as granulosa cells, and actively synthesize steroid hormones when provided with appropriate substrates (Channing 1966), a similar rôle in vivo should not be overlooked. Not only may these cells synthesize steroids, but also they may retain an ability to secrete diverse peptides, thus influencing both the process of fertilization and the first stages of embryonic development. There is evidence, moreover, that isolated cumulus cells readily form pyruvate in the presence of glucose and lactate (Leese and Barton 1985), suggesting a specific rôle in embryonic nutrition.

Fertilizin-Antifertilizin Reactions at the Egg Surface

Evidence that a fertilizin-antifertilizin reaction is a critical step preceding the fertilization process of certain invertebrates (e.g. the sea-urchin *Arbacia*) is well accepted (Lillie 1912; Austin 1961, 1965). A factor emanating from the egg coat (fertilizin) reacts with molecules on the sperm surface (antifertilizin) leading to agglutination and seemingly stimulating the acrosome reaction. Fertilizin is a glycoprotein and diffuses from the jelly coat of sea-urchin eggs. Sea water in which the eggs have been allowed to stand has the ability to agglutinate spermatozoa, since the liberated fertilizin reacts with a sperm surface component termed antifertilizin − an acidic protein. Fertilizin-antifertilizin reactions are best regarded as molecular preliminaries to fertilization, underlying the recognition and attachment of homologous spermatozoa to the egg surface.

The extent to which mammalian eggs possess a "fertilizin" associated with the surface membranes which requires a period of exposure to a putative antifertilizin in the tubal fluids before spermatozoa can penetrate the zona pellucida remains uncertain. In a classical paper on mammals, Bishop and Tyler (1956) claimed analogies with the fertilizin reaction in invertebrates, suggesting that a substance diffused from the zona pellucida to react with spermatozoa and increase their tendency to autoagglutinate. However, the in vitro studies of Thibault and Dauzier (1960) offered a rather different viewpoint. They noted that washing rabbit eggs in a physiological medium before addition of a sperm suspension resulted in a significantly higher incidence of in vitro fertilization, inferring that a substance was removed from or neutralized on the egg surface in the Fallopian tube. A subsequent study in the rabbit failed to find any evidence for a specific fertilizin emanating from the egg surface which might be involved in initiating the acrosome reaction, thereby facilitating a higher incidence of sperm penetration (Overstreet and Bedford 1975). In longer essays, Bedford (1982, 1983) argues from various points of view against the existence of a fertilizin in mammals but, as already remarked, there may be time during egg transport to the site of fertilization for subtle changes in the egg surface corresponding to the influence of washing in vitro.

Binding Sites for Spermatozoa on the Zona Pellucida

Whereas the evidence for fertilizin-antifertilizin reactions in mammalian eggs remains equivocal, there are now compelling observations to indicate development of specific receptor or binding sites for spermatozoa on the surface of the zona pellucida (Gwatkin 1977; Yanagimachi 1981; Dunbar 1983; Hartmann 1983). At first reading, this may appear surprising, since the zona pellucida is an accessory acellular layer quite distinct from the plasma membrane surrounding the cytoplasm of the egg. Moreover, the outermost portion of the zona is sufficiently loose, spongy and uneven (Fig. V.3) for it to arrest motile spermatozoa − in which case other more specific forms of contact interaction might seem unnecessary for penetration to proceed. Nonetheless, the zona is known to be a biochemically complex glycoprotein matrix (Dunbar 1983; Sacco, Yurewicz and Subramanian 1986), and its surface develops specific sugar residues which can be demonstrated as receptors by their ability to bind lectins, a group of plant and invertebrate proteins (Nicolson and Yanagimachi 1972; Nicolson et al. 1977). These sugar residues on the zona have a special affinity for carbohydrate moieties on the anterior portion of the sperm head and in one sense, therefore, chemical binding of the sperm head to the zona pellucida can be viewed as the first functional contact between the two gametes. Species-specificity of fertilization is thought to reside in part in the zona binding and subsequent penetration mechanisms, there being only extremely rare reports of inter-specific sperm penetration after experimental insemination in vivo or in vitro when the egg is intact (see Dickmann 1962).

The binding reaction between male and female gametes must be remarkably avid, because viable spermatozoa reaching the surface of the zona pellucida are usually exhibiting a high degree of progressive motility and might thus be expected to lose contact with the surface of a sphere. Perhaps, after all, a degree

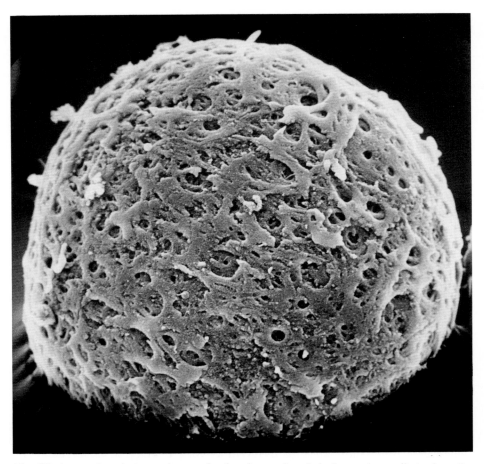

Fig. V.3. A scanning electron micrograph of a pig egg denuded of granulosa cells to show the uneven, loose and spongy outer surface of the zona pellucida. Such a surface may facilitate contact reactions between the spermatozoon and egg. (Courtesy of Dr. J. E. Fléchon; after Hunter 1980)

of physical entrapment of the sperm head between corona cells or by the uneven surface of the zona is important whilst the binding reaction is being initiated. Two related points are worth mentioning. First, there are not preferential regions on the zona for sperm attachment: rather, the binding sites are apparently distributed over the complete surface of the sphere. This seems an essential prerequisite since the portion of the sphere first contacted by a competent spermatozoon is probably at random. Second, formation of the zona pellucida is not completed in small Graafian follicles; on the contrary, material is added to the zona right up to the moment of ovulation in several species of mammal and especially the large farm animals (Fig. V.4), suggesting a comprehensive distribution of binding sites.

Various research groups have devoted specific attention to the "zona binding" of spermatozoa, and have especially studied gametes of the golden hamster under

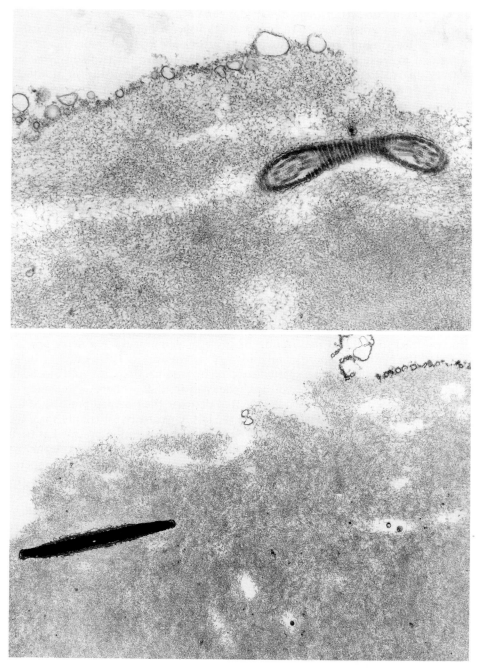

Fig. V.4. Thin sections of a portion of the zona pellucida of a pig egg to show the spongy and porous nature of the outer surface, suggesting a final peri-ovulatory addition of zona material. (Courtesy of Dr. D. Szöllösi)

in vitro conditions (Hartmann 1983); the zona binding is suggested to occur in two stages: an initial weak attachment of the sperm head to the zona surface followed by a more compact bonding as actual penetration commences. It is uncertain to what extent the same reaction is involved in the two somewhat arbitrary stages. Bonding might simply represent a more advanced form of attachment. Nor is it clear precisely how (or when) a distinction is drawn between attachment and the initial stages of zona penetration − that is entry of the tip of the sperm head into the substance of the zona. As stated succinctly by Yanagimachi (1981): "To understand the molecular mechanisms of sperm-zona interaction, we must fully understand the chemical characteristics of both the sperm surface and the zona pellucida."

Capacitation of Spermatozoa, the Acrosome Reaction and Hyperactivation

The capacitation of mammalian spermatozoa has been a favoured topic of research since the phenomenon was first specifically identified by Austin (1951) and Chang (1951). These two biologists were working independently with different species on different continents, and examining chronological aspects of the early stages of fertilization in rats and rabbits. Although their experimental techniques were not the same, the essential finding in both studies was that a delay appeared necessary between the time of spermatozoa entering the female reproductive tract and actual penetration of the egg membranes − a delay not accounted for by the transport or migration of spermatozoa. This period of delay was regarded as one in which spermatozoa underwent physiological and biochemical changes and finally achieved the capacity to fertilize − hence the choice of the word "capacitation" by Austin (1952) to describe the ripening process. Many relevant observations have been made in the last 35 years, but it is only since the early 1970s that an especially critical feature of the capacitation process has been recognized: i.e. a change in the sperm surface that enables a specific influx of calcium ions which, in turn, lead to alterations in the configuration of the sperm head membranes (Yanagimachi and Usui 1974). The membrane changes on the sperm head, culminating in the acrosome reaction (Fig. V.5), are paralleled by changes in motility generated in the mid-piece mitochondria and expressed in movements of the flagellum. These head and mid-piece modifications together confer on the sperm cell an ability to penetrate to the egg surface.

A distinction has been drawn between capacitation and these subsequent changes: indeed, as long ago as 1970, capacitation was being viewed as a physiological change or series of changes in the sperm cell that precedes and permits a coordinated acrosome reaction (Austin 1967; Bedford 1970; Chang and Hunter 1975). As inferred, a more recent definition would also include reference to the closely coincident or simultaneous changes in motility. Descriptive detail abounds on the process of capacitation, derived from studies in vivo and in vitro in a number of mammalian species, so it is perhaps worth attempting to place the process in a biological perspective before highlighting some of its many facets.

Because spermatozoa can be viewed as fragile and potentially unstable cells with very limited reserves of energy (negligible cytoplasm) and a finite comple-

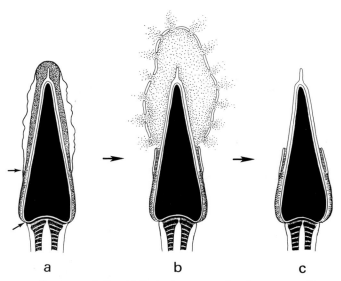

Fig. V.5. Diagrammatic representation of the acrosome reaction in a mammalian spermatozoon to show the multiple fusions between the outer acrosomal membrane and overlying plasma membrane. This vesiculation on the anterior portion of the sperm head permits release of lytic enzymes. (Courtesy of Dr. J. M. Bedford; after Hunter 1980)

ment of lytic enzymes in the acrosome, it seems vital to preserve these components until spermatozoa are in the vicinity of an egg at the site of fertilization. Exhaustion of spermatozoa in the lower reaches of the female tract or loss of their complement of acrosomal enzymes at this point would render the cells non-functional. There is thus a *physical* requirement to maintain sperm integrity until entry into the Fallopian tubes. There is also a *chronological* requirement, in the sense that mating may precede ovulation by many hours in laboratory species or several days in pigs, dogs, horses and primates; with such timescales, there would be no conceivable value in achieving full capacitation and hyperactive motility shortly after semen deposition in the female tract, for spermatozoa would simply not remain functional until ovulation. Nonetheless, inspection of many of the established texts suggests that capacitation is a process requiring only a small number of hours irrespective of the time of mating before ovulation (Table V.1). Indeed, Fraser (1984) stated that capacitation is "a time-dependent phenomenon, with the absolute period being species-dependent".

A frequently-offered explanation for the continued availability of capacitated cells many hours after mating invokes the heterogeneous condition within the vast numbers of spermatozoa in the ejaculate (i.e. immature, ripe, moribund and dead). Small populations of initially immature cells would pass sequentially up a "curve of ripening" during the pre-ovulatory interval (Fig. V.6), so that there should always be some competent cells available at the time of ovulation (see Bedford 1970, 1982, 1983; Dziuk 1970; Chang and Hunter 1975). If this interpretation retains any validity in the light of the previous remarks, then at best it seems likely to represent a secondary mechanism. Certainly, it is important to note that

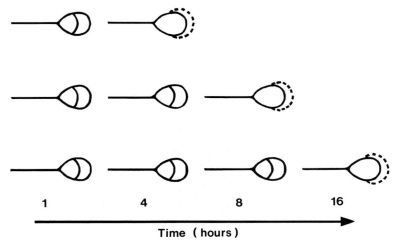

Fig. V.6. 'Curves' of sperm maturation to correspond with the hypothetical and sequential achievement of capacitation in different populations of spermatozoa within the ejaculate. (Redrawn after Bedford 1982)

Table V.1. Approximate values for the capacitation time of mammalian spermatozoa in vivo available in a number of established textbooks. (After Hunter 1987b)

Species	Interval (h)
Sheep	1 – 1.5
Pig	2 – 3
Rat	2 – 3
Hamster	3 – 4
Cow	4 – 5
Rabbit	5 – 6

classical studies of capacitation, both in vivo and in vitro, have invariably been performed in a "post-ovulatory environment" − that is in the presence of eggs and/or their investments, and this may have contributed to a proposed misunderstanding (Hunter 1986, 1987a, b). In fact, tables of specific (fixed) values for the timing of capacitation may not be biologically meaningful.

A more realistic interpretation in eutherian mammals may be associated with the pre-ovulatory physiology of the Fallopian tubes and the endocrine activity of the Graafian follicle − the structure that also releases the female gamete. Here, the suggestion derived from studies of sperm distribution in the tubes of farm animals is that physiological control of capacitation is obtained via a peri-ovulatory transfer of follicular hormones to the tissues of the distal isthmus, the region in which viable spermatozoa are largely arrested and stored in the pre-ovulatory interval (see Chap. IV). For example, ovarian follicular hormones such as prostaglandins increase dramatically in concentration just before ovulation

(Tsafriri et al. 1972; LeMaire et al. 1973; Ainsworth et al. 1975; Hunter and Poyser 1985) and may act directly on the sperm cells (Kelly 1981; Aitken and Kelly 1985), or a local transfer of ovarian hormones may prompt the release of catecholamines into the lumen of the tube. Sperm cells are known to have receptors for such molecules, and to respond to appropriate catecholamine stimulation by changes in motility and membrane configuration (Bavister and Teichman 1976; Bavister, Chen and Fu 1979; Meizel et al. 1980). The underlying strategy would be to achieve full capacitation and hyperactivation of spermatozoa just as the egg was being released into the tract rather than many hours before ovulation. Thus, the endocrine activity of the Graafian follicle would serve to integrate the final phase of sperm transport to the site of fertilization with the process of capacitation (Hunter and Nichol 1983; Hunter 1984, 1986, 1987 a, b).

In brief, therefore, this hypothesis proposes that instead of components derived from eggs being the principal stimulus to the final maturational changes in spermatozoa — as in marine invertebrates — eutherian mammals may have developed a strategy of ovarian endocrine regulation of events within both the reproductive tract and, indirectly, the sperm cell itself. Such an interpretation would argue strongly that capacitation times do not correspond to the rigid figures presented in Table V.1. Evidence obtained in vitro is of considerable interest and may support such estimates, but the evidence probably says more about the potential rate of induction of capacitation in unselected populations of spermatozoa in relatively large volumes of culture fluid than it clarifies the subtle and remarkable control mechanisms in the specific fluids of the Fallopian tubes. Completion of capacitation may best be thought of as an ovulation-related event (Hunter 1987 a, b).

Details of the capacitation process have been discussed comprehensively in numerous reviews (Bedford 1970, 1983; Austin 1975; Chang and Hunter 1975; Chang et al. 1977; Gwatkin 1977; Rogers 1978; Yanagimachi 1981; Fraser 1984; Hunter 1987 a, b); since the following treatment is relatively brief and selective, readers are referred to these reviews for a full coverage. An essential preliminary to capacitation is liberation of the sperm cell from the seminal plasma in which it is suspended at ejaculation. It must be recalled that seminal plasma represents the secretions not only of the male accessory glands but also of the epididymal duct, in which portion of the tract spermatozoa may have been stabilized and stored for as long as 2 to 3 weeks. As already discussed in Chapter IV, most of the seminal plasma is removed from the sperm cells during passage through the cervix and/or the utero-tubal junction. Components of the seminal plasma that are more intimately associated with the sperm surface are progressively leached or metabolized as spermatozoa are resuspended in female genital tract fluids during passage towards the site of fertilization. Chang (1957) demonstrated that spermatozoa presumed to be capacitated could be decapacitated by addition of seminal plasma, such spermatozoa then requiring a further period in the female tract before they were capable of fertilization. Although this experiment indicates the necessity of removing seminal plasma, it is doubtful whether fully-capacitated spermatozoa could be rendered functionally immature once more by exposure to male fluids: irreversible changes in the cell would almost certainly have occurred.

Several studies have now demonstrated a sequential and coordinated action of compartments within the female tract on the rate of capacitation, in the sense that spermatozoa passing through the uterus into the Fallopian tubes are capacitated more rapidly than a suspension of ejaculated spermatozoa deposited directly into the tubes (Adams and Chang 1962; Bedford 1968a; Hunter 1969; Hunter and Hall 1974). The principal explanation here may be that seminal plasma would initially still be in contact with spermatozoa deposited into the tubes, whereas it would have been largely removed during exposure to preceding compartments of the female tract. However, a population of spermatozoa introduced surgically into the tubes would undoubtedly be numerically greater than that entering it after mating, and there is evidence from rabbits (Bedford 1968a) and pigs (unpublished observations) that the Fallopian tubes may capacitate only limited numbers of spermatozoa at one time. Subtle biochemical forms of synergism could therefore be at play.

Escape from the seminal plasma may be associated with loss of specific inhibitors of acrosomal enzymes and perhaps also of mitochondrial enzymes. Such inhibitors may be acquired by the cells during passage through the epididymis, as appears to be the case in boar spermatozoa (Hunter, Holtz and Herrmann 1978). Increased metabolic activity can be detected as capacitation proceeds, underlying in part the increase in sperm motility. In the golden hamster, the effects of the tubal environment on stimulated metabolism and achievement of the capacitated state are so distinct that they can be monitored in vivo through the thin wall of the ampulla in terms of a massive increase in sperm motility (Yanagimachi 1970); this has also been reported for the mouse (Suarez 1987). Influences of female tract fluids on sperm metabolism and capacitation have been reviewed elsewhere (Bedford 1970; Chang and Hunter 1975; Gwatkin 1977; Harrison 1977; Fraser 1984). Gonadal fluids in the form of ovarian follicular fluid may also contribute to capacitation in species such as the golden hamster (Yanagimachi 1969a, b). If changes promoted by follicular fluid are part of the spontaneous in vivo mechanism of capacitation, then the inference is that capacitation is not completed until the time of ovulation – rather than 3–4 hours after mating in hamsters.

A major focus of the capacitation process appears to be an alteration in the lipid and/or protein domains in the plasma membrane such that Ca^{2+} can enter the cell and lead to changes expressed as both the acrosome reaction and a dramatically enhanced swimming activity termed hyperactivation (Yanagimachi 1981). The calcium binding protein, calmodulin, may serve to regulate calcium fluxes. There is a view that, given an appropriate fluid environment, progressive destabilization changes in the plasma membrane may be largely under intrinsic control of the sperm cell rather than promoted by female tract fluids (Bavister 1969; Austin and Bavister 1974). If this were so, it would explain why capacitation can be obtained in various in vitro systems containing suitable concentrations of electrolytes, metabolic substrates, and serum albumin as the sole added protein. On the other hand, this view would reduce or deny the importance of specific ovarian influences on the rate of capacitation which, as inferred above, may be critical in species with a relatively long interval between the gonadotrophin surge and ovulation (e.g. sheep, cow, pig, horse).

As already indicated, the acrosome reaction occurs as a consequence of capacitation. It is thus a distinct process and stems from a destabilization of the plasma membrane around the sperm head. But the extent to which the mammalian egg or its investments promote the acrosome reaction has still to be resolved in the physiological situation (Szollosi and Hunter 1978). Analogies with the events of fertilization in marine invertebrates are attractive, and there is striking evidence from fertilization of human eggs in vitro, but these do not permit statements on the site and causes of the reaction in vivo. However, relevant observations are available for the large farm species (Crozet and Dumont 1984; Herz et al. 1985) and the hamster (Cummins and Yanagimachi 1982).

The acrosome reaction of mammalian spermatozoa has been widely described as a vesiculation reaction involving extensive points of fusion between the plasma membrane and the underlying outer acrosomal membrane (Austin 1963b, 1965; Barros et al. 1967; Bedford 1967, 1968b, 1970; Piko 1969; Yanagimachi and Noda 1970a). In this regard, one crucial function of capacitation may be to arrange a specific distance between these two membranes so that vesiculation can proceed. Actual vesicles are formed by the rounding off of small portions of plasmalemma and the outer acrosomal membrane, and this spreads in three dimensions over the sperm head (Fig. V.5). Formation of such vesicles was initially thought to be spontaneous and simultaneous once capacitation had occurred, but Fléchon (1985), using the technique of freeze-fracture coupled with scanning electron microscopy, has demonstrated that the acrosome reaction in ram and/or rabbit spermatozoa proceeds anteriorly from the region of the equatorial segment (Fig. V.7).

Although the acrosome reaction may promote specific changes in surface characteristics of the posterior part of the sperm head (Yanagimachi 1981; Bedford 1983), its primary function is still seen as a means of releasing proteolytic enzymes from the acrosomal sac to facilitate penetration of the egg investments. As a modified lysosome, the acrosome is derived from the Golgi apparatus and contains a large number of enzymes, but the two invariably considered in essays on fertilization are hyaluronidase and acrosin. As already noted, hyaluronidase acts to depolymerize hyaluronic acid, the cement substance between the cumulus cells, thereby giving the sperm access to the surface of the zona pellucida. Acrosin, a trypsin-like enzyme, formerly referred to as the hypothetical zona lysin (see Austin 1961), may digest a pathway through the substance of the zona (Fig. V.8), but controversy exists on this point and also on whether acrosin is largely membrane-bound or freely released during the acrosome reaction. One argument for the enzyme being membrane-bound is that the penetrating sperm leaves a discrete slit in the zona − which would not be expected if a lytic enzyme was freely diffusing from the sperm head. What is seemingly not controversial is the proposition that acrosin exists in a zymogen form, proacrosin, which needs to be activated to acrosin during the acrosome reaction (Hartree 1977; Brown and Harrison 1978; Harrison, Fléchon and Brown 1982). Indeed, such enzymatic modification may itself be involved in promoting the membrane fusion reaction (Meizel 1984). The contribution, if any, of the remaining complement of acrosomal enzymes in the events of fertilization awaits systematic study.

A dramatic increase in the rate of flagellar beat causing a much enhanced progressive motility is a second functional expression of capacitation in many mam-

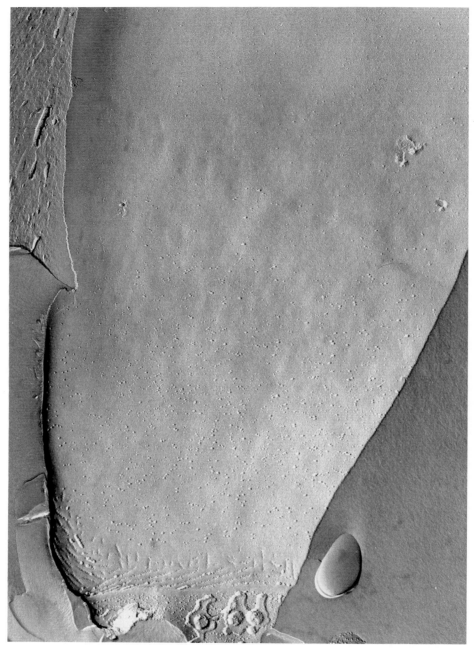

Fig. V.7. Freeze-fracture preparation of a sperm head in which the acrosome reaction proceeds in an anterior direction from the equatorial segment. (Courtesy of Dr. J. E. Fléchon)

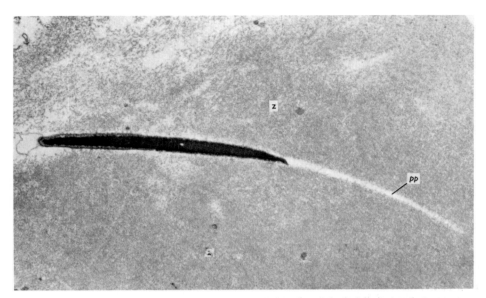

Fig. V.8. Segment of the zona pellucida of a newly-fertilized pig egg to show a portion of the sperm penetration pathway in the substance of the zona. (After Szöllösi and Hunter 1973)

mals. The configuration of the beat of the tail changes to an "S-shaped" or whiplash condition (Yanagimachi 1981; Fraser 1984), and the propulsive forces so engendered have been calculated (Katz, Yanagimachi and Dresdner 1978). Hyperactivation is the term now used to describe this form of motility in capacitated spermatozoa which, clearly, must soon lead to exhaustion. As already emphasized, an influx of Ca^{2+} ions is important in the hyperactivation phenomenon, as is also a modification to the ATP status of the cell — events presumably focussed on the mid-piece mitochondria.

In recent years, there has been a trend amongst biologists studying fertilization, (as distinct from biochemists), to question the extent to which acrosomal enzymes are involved in penetration of the zona pellucida. Two of the considerations have already been discussed: the discrete sperm penetration slit and the condition of hyperactivated whiplash motility with its attendant propulsive forces. A third is the physical condition of the sperm head expressed as a very considerable rigidity. During the phase of maturation in the epididymis, the nuclear chromatin of testicular spermatozoa is stabilized and more tightly condensed by the formation of disulphide bridges between adjacent strands of chromatin (Calvin and Bedford 1971). Upon incorporation into the vitellus at the beginning of fertilization, a breakdown of the bonding and a decondensation of the sperm nuclear chromatin are two of the earliest detectable changes. This could suggest that sperm head rigidity is important during transit in the female tract and/or for actual penetration of the egg investments, especially for traversing the zona pellucida. Arising from these observations is the possibility that sperm penetration of the zona is primarily mechanical due to the incisive nature of a rigid head linked to the propulsive force of a hyperactivated flagellum (Bedford 1982, 1983). Until conclusive evidence is presented for one view or the other, a more balanced position would be to invoke a rôle for both the enzymatic and mechanical components (see Hunter 1979, 1980).

Penetration of the Zona and Fusion of Gametes

Irrespective of the precise mechanisms of penetration of the zona pellucida, the form of the pathway through this egg coat has been described repeatedly since the studies of Dickmann (1964, 1965). Spermatozoa commence penetration from a tangential position, effectively lying flat on the outer surface of the zona and then they pursue a pathway of ever increasing steepness towards the perivitelline space. The angle of passage through the innermost portion of the zona may approach the perpendicular. Explanations for the form of the pathway could be related in part to differing densities within the substance of the zona. As indicated, the flagellum remains hyperactive during penetration, but once the sperm head has entered the perivitelline space and made a functional contact with the egg plasmalemma, movement of the flagellum rapidly ceases. Incorporation of the sperm midpiece and tail into the vitellus then becomes dependent upon the egg surface rather than on forces in the sperm itself.

The actual mode of sperm incorporation has been studied by transmission and scanning electron microscopy, using various models and/or degrees of polyspermy to increase the chance of obtaining gametes at the appropriate stage of interaction (Bedford 1968b, 1972; Yanagimachi and Noda 1970b). A consistent

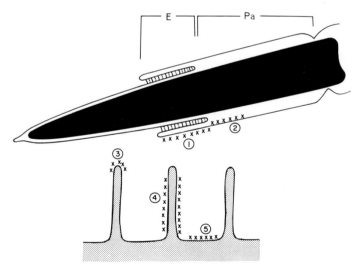

Fig. V.9. Diagrammatic representation of a mammalian sperm head at the moment preceding incorporation by the egg plasmalemma. The equatorial segment (*E*) and post-acrosomal region (*Pa*) are designated, whereas the numbers refer to characteristics of the surface membranes. (Courtesy of Dr. R. Yanagimachi)

96

finding in all these studies is that the anterior portion of the sperm head, now bounded by the inner acrosomal membrane, is not the portion to be incorporated first. Rather, it is the region of the equatorial segment itself that is apparently attractive to the egg plasmalemma, and it is here that the first signs of fusion are noted (Fig. V.9). One current interpretation is that the events of the acrosome reaction and the resultant establishment of membrane continuity in the equatorial region lead to a change in surface charge in this region and around the postnuclear cap (Yanagimachi 1981). The underlying molecular events have yet to be clarified, but freeze-fracture studies indicate considerable intra- and sub-membranous movement of particulate matter at this stage (Fléchon 1985).

Whilst the ultrastructural studies of Piko and Tyler (1964) suggested that vitelline incorporation of the sperm head was closely analogous to phagocytosis, involving formation of a vesicle, it is now widely accepted that sperm-egg fusion in mammals leads to the constitution of a mosaic membrane at the egg surface due to insertion of the sperm plasma membrane. Such membranous modification in mammals was first indicated by the study of Szollosi and Ris (1961), and could have significant immunological implications.

Activation of the Egg and Block to Polyspermy

The responses of the egg (secondary oocyte) to functional contact with a sperm cell are generally referred to as *activation*. The best-known response to sperm penetration is completion of the second meiotic division, but there are also modifications to the cortical organelles, the cytoskeleton, and various biosynthetic processes.

As discussed in Chapter IV, a primary oocyte resumes meiosis in response to the gonadotrophin surge shortly before ovulation and enters the Fallopian tube as a secondary oocyte. The fertilizing spermatozoon stimulates completion of the meiotic division, with extrusion of the second polar body and arrangement of the haploid group of female chromosomes as a compact lump of chromatin. The latter becomes more diffusely arranged and initiates formation of a pronuclear membrane (Fig. V.10). The second polar body is a redundant structure, has few organelles, and fails to form a nuclear membrane. The male pronucleus is invariably larger than its female counterpart, seemingly because male elements commence pronuclear formation whilst anaphase and telophase of the second meiotic division are still proceeding. In fact, immediately upon incorporation at the vitelline surface, the chromatin of the sperm head starts a characteristic decondensation and swelling, this originating in the region beneath the postnuclear cap (Bedford 1968b; Yanagimachi and Noda 1970b). A new unit membrane appears around the chromatin as the pronucleus forms. Male and female pronuclei migrate towards the centre of the egg.

Completion of the second meiotic division is essential if the penetrated egg is to avoid an abnormal chromosome constitution. Equally important is instigation of a defence mechanism against multiple sperm penetration into the vitellus: this is referred to as the block to polyspermy (Austin and Braden 1953; Braden, Austin and David 1954). Although it has an electrical component (see Miyazaki

Fig. V.10. Thin section of a recently-activated pig egg to show the microtubules of the spindle apparatus at the completion of the second meiotic division. Note that a membrane fails to form around the chromatin of the second polar body. (After Szöllösi and Hunter 1973)

and Igusa 1981), its basis resides principally in an action of the contents of the cortical granules (Austin 1956; Szollosi 1962, 1967; Fléchon 1970; Gulyas 1980), the latter being vesicles of approximately 1 μm diameter located immediately beneath the plasma membrane in the newly-ovulated egg. They are derived from the Golgi apparatus and migrate towards the egg surface, their peripheral disposition being especially noticeable after the pre-ovulatory gonadotrophin surge.

Activation of the cortical granules leads to a process of membrane fusion between the organelles and the overlying plasmalemma, such that escape ports are formed and the contents of individual vesicles are released into the perivitelline space. Since this cortical reaction is also a membrane vesiculation reaction, Ca^{2+} ions are presumed to play a key rôle in the process of exocytosis. The contents of the granules are thought to be enzymatic in nature (Gwatkin and Williams 1974; Guraya 1983), and diffuse across the still quite narrow and fluid-filled perivitelline space. In most mammalian species so far studied, the contents act on the proteins or glycoproteins of the zona pellucida to prevent complete penetration by further spermatozoa. This enzymatic action appears to be restricted to the innermost region of the zona, since accessory spermatozoa can continue to penetrate outer portions but cannot traverse completely across the zona. Structural changes in the zona in response to cortical granule material have so far not been distinguished by microscopy. On the other hand, cortical granule material obtained from unfertilized eggs by sonication and centrifugation can be used to treat the zona pellucida of fresh unfertilized eggs, and can render them impenetrable to capacitated spermatozoa (Barros and Yanagimachi 1971; Gwatkin et al. 1973). Whilst the precise mechanism of action of cortical granule material is not fully understood, it may antagonize a lytic rôle of acrosomal enzymes within the zona, thus preventing sperm penetration.

Formation of a block in the zona pellucida is the most common form of defence against polyspermy, but this block does not exist in rabbit eggs. Here, protection against polyspermic penetration is at the level of the vitelline membrane, so large numbers of spermatozoa may accumulate in the perivitelline space (Braden et al. 1954).

Other forms of activation in the newly-penetrated egg include (1) contraction of the vitellus, a change which infers modifications to the cytoskeleton, perhaps involving the proteins actin and tubulin; and (2) initiation and/or reprogramming of synthetic processes in the cytoplasm and the pronuclei themselves. A clear demonstration of the latter is found in both the male and female pronuclei which replicate their DNA complement during migration to the centre of the egg (Oprescu and Thibault 1965; Szollosi 1966). DNA replication is required during this migratory phase because a resting nucleus is not formed when the pronuclei meet: instead the chromosomes condense at metaphase of the first mitotic division and proceed directly to anaphase and telophase followed by the cleavage division.

Synchrony of Penetration and Cleavage

In polyovular species such as rabbits, rodents and pigs, an estimate can be made of the time required for all the eggs released at one mating period to be penetrated and activated (e.g. Austin and Braden 1954). However, actual observations on the stages of egg development do not distinguish between a spread in the time of ovulation of individual Graafian follicles and variation in the time needed for competent spermatozoa to reach the surface of the zona pellucida. As a general remark, though, spontaneous ovulation in a group of follicles appears closely synchronous, presumably because mature follicles are responding to the same surge of gonadotrophic hormone. In part due to this synchrony, freshly-released eggs in rabbits, hamsters and pigs usually aggregate within a cumulus plug (Hancock 1961; Harper 1961) and proceed to the site of fertilization in that condition.

Actual observation of the process of ovulation in domestic pigs suggested periods of between 1 – 5 h for its completion (Pitkjanen 1958; Ito, Kudo and Niwa 1959; Betteridge and Raeside 1962), but almost certainly the observational techniques led to a protracted sequence of changes. On the other hand, pig ovaries exposed through mid-ventral laparotomy just before the anticipated time of ovulation have been noted to undergo follicular collapse within 1 – 5 min (unpublished observations). Likewise, in several hundred pairs of pig ovaries examined at autopsy around the expected time of ovulation, either pendulous follicles on the verge of collapse or recent ovulations were seen; seldom ($<15\%$) were the two conditions present together in the same ovaries, suggesting a closely synchronous ovulation of the 10–18 mature follicles. Observations in laboratory species using an abdominal window technique (Blandau 1955) have also suggested that ovulation is relatively closely synchronized. This synchrony may not necessarily be found in women whose ovaries have been stimulated with gonadotrophins to enhance the number of pre-ovulatory Graafian follicles; it is certainly not found in superovulated cows (Rowson 1951; Betteridge 1977).

As to the timing of sperm penetration of individual eggs, there is evidence of a spread here, possibly extending to several hours or more (Braden and Austin 1954; Dziuk 1965; Gates 1965; Thibault 1967; Hunter 1972). Explanations for such a spread would include the small number of spermatozoa initially reaching the site of fertilization, and the physical arrangement of the cumulus mass within the lumen of the tube. The rate of penetration of the cumulus investment to the surface of the zona would probably vary between oocytes, and the time required for actual penetration of the zona could introduce further variation (Yanagimachi 1981). Together, these influences might introduce a spread of a few hours into the timing of activation of all eggs in a group, and this would then be seen in the tim-

Table V.2. Chronology of preimplantation development in five mammals, and stage of embryonic passage into the uterus from the Fallopian tubes. Preovulatory mating or insemination has been assumed, and the stages are therefore presented as times after ovulation or sperm penetration of the eggs. (Modified after Hunter 1980)

Species	First cleavage (h)	Second cleavage (h)	Entry into uterus	
			Time (h)	Stage of development
Rabbit	20 – 24	25 – 32	60 – 76	Morula
Cow	20 – 24	32 – 36	72 – 84	8 – 16 cells
Sheep	16 – 18	28 – 30	66 – 72	8 – 16 cells
Pig	14 – 16	20 – 24	46 – 48	4 cells
Horse	24	30 – 36	140 – 144[a]	Blastocysts

[a] Approximate figure for fertilized eggs; unfertilized eggs remain and degenerate in the Fallopian tube.

ing of the first cleavage division. Certainly, single-cell fertilized eggs and two-cell-ed embryos can be recovered from the Fallopian tubes of the same animal after natural mating and spontaneous ovulation. A greater range in cleavage stages of embryos recovered from the same Fallopian tube or uterine horn may be noted as development proceeds, but this may be at least as much an expression of internally programmed rates of cleavage as a carry-over effect of variation in the timing of sperm penetration.

Some generalized figures for the rates of early embryonic development are presented in Table V.2.

Abnormalities of Fertilization

Abnormalities arising around the time of fertilization are of considerable significance in domestic livestock, and are presumed also to be so in women, since they constitute a major source of embryonic loss. In the absence of any specific pathology, two of the principal abnormalities of fertilization are associated with post-ovulatory ageing of eggs in the Fallopian tubes before sperm penetration; these are (1) digyny (retention of the second polar body) and (2) polyspermy (penetration of the vitellus by more than one spermatozoon). They arise due to malfunction of the egg organelles. Disorganisation of the microtubules of the meiotic spindle by a form of rotation may cause both groups of female chromosomes to remain within the vitellus at telophase. If pronuclei then develop, there is the potential for formation of a triploid at syngamy. Other forms of spindle anomaly arising at the time of activation, such as non-disjunction of one or more pairs of chromosomes, could lead to disturbance of haploid complements and thereby to aneuploids.

The cortical granules also undergo modification with post-ovulatory ageing, although this is initially one of location rather than structure. These vesicles wander from just beneath the vitelline membrane, a position assumed at the time of ovulation (see above), and become scattered in deeper portions of the cortex (Szollosi 1975). Such a distribution prevents the liberation of their contents by membrane fusion with the plasmalemma at the time of egg activation, and a block to polyspermy is therefore not established. In fact, the polyspermic condition may be enhanced in this situation of ageing, since regulation by the isthmus of sperm numbers passing to the site of fertilization will be progressively less effective as the post-ovulatory interval increases. The incidence of polyspermy rises both with ageing of the eggs (Hancock 1959; Thibault 1959; Hunter 1967a) and with elevated numbers of spermatozoa at the site of fertilization (Polge, Salamon and Wilmut 1970; Hunter 1973). Viable lifespans of the gametes have been referred to in Table IV.6 and the consequences of post-ovulatory ageing on fertilization and embryonic loss are illustrated in Table V.3.

The fertilizing spermatozoon itself may give rise to chromosomal anomalies, since diploid spermatozoa have been recorded and, on occasions, are known to be capable of penetration (Fechheimer and Beatty 1974; Mortimer 1978). Irrespective of the route of their formation, triploid or polyploid embryos are invariably lethal conditions in mammals (Beatty 1957, 1961; Austin 1963a, 1965;

Table V.3. The influence of post-ovulatory ageing of pig eggs in the Fallopian tubes on fertilization and on embryonic survival at 25 days after insemination. Eighteen animals were inseminated in each group. (Compiled from Hunter 1967a, b)

Estimated age of eggs at fertilization	Eggs fertilized normally % ± (S.E.)	Viable embryos at 25 days	
		% survival ± (S.E.)	Mean number
0	90.8 (4.5)	87.9 (2.9)	12.0
4	92.1 (2.7)	72.9 (14.9)	11.7
8	94.6 (2.3)	60.5 (13.2)	8.7
12	70.3 (7.8)	53.3 (15.7)	6.8
16	48.3 (8.4)	27.9 (14.5)	4.8
20	50.9 (7.5)	32.3 (15.2)	5.0

Thibault 1972), and normally die at an early stage of development (Bomsel-Helmreich 1965, 1971). Aneuploids, on the other hand, may progress to an advanced stage of gestation or to term. Suspected abnormalities of the chromosome complements in young embryos can be revealed by karyotyping, a classical microscopic technique applied to metaphase spreads. As a more recent advance, abnormalities of the sperm's karyotype may be revealed using the zona-free hamster egg to incorporate the gamete and decondense the nuclear chromatin (see Chap. IX).

This brief discussion has focussed on gross genetic errors, i.e. at the level of the chromosomes, that can be revealed by appropriate microscopic techniques. However, there is little doubt that more subtle errors involving individual genes or groups of genes are also associated with early death of embryos. Indeed, an elegant thesis linking putative genetic errors in the gametes with subsequent embryonic death was published by Bishop (1964).

Fate of Redundant Gametes

Spermatozoa entering the female genital tract suffer a number of fates (Cohen and Adeghe 1987). A privileged few gain the site of fertilization and become associated with the egg investments, leading to entrapment in the zona pellucida (as a result of the block to polyspermy) or incorporation in follicular cells. Whilst these processes may account for several hundred spermatozoa, this represents only a tiny proportion of the cells in the ejaculate. The great majority of spermatozoa are lost from the female tract by leakage or voiding within a short interval of mating; of those gaining the uterus, significant numbers are phagocytosed by polymorphonuclear leucocytes (Austin 1964; Lovell and Getty 1968). There is controversy as to whether phagocytosis of spermatozoa occurs in the Fallopian tubes, at least while the embryo resides there. Excluding situations of infection, this is thought not to be the case in farm animals but there is some evidence to the contrary in rodents (Austin 1959). Considerable numbers of spermatozoa may traverse the length of the Fallopian tubes and enter the peritoneal cavity; this has long been known to occur in the absence of an ovarian bursa, and in women has permitted a clinical test of tubal patency. Samples of peritoneal fluid, preferably aspirated from the pouch of Douglas, are monitored for their sperm populations (e.g. Templeton and Mortimer 1980). Finally, there is some evidence that those spermatozoa in the isthmus of the tubes no longer capable of interacting with the egg are moved passively into the uterus at the time of embryo descent, where they may be incorporated by elements of the proliferating trophoblast (Hunter 1978).

Unfertilized eggs show a progressive degeneration of the cytoplasm, which may include a phase of fragmentation (Dziuk 1960). They also show a gradual softening and distortion of the zona pellucida, but this remains around the egg whilst in the Fallopian tube. In most mammals so far examined, unfertilized eggs enter the uterus on the same time-scale as fertilized eggs, but this is not so in horses (van Niekerk and Gerneke 1966; Betteridge and Mitchell 1974) and certain species of bat (Rasweiler 1972). Here, the unfertilized eggs are retained for prolonged periods in the tube, even whilst fertilized eggs progress past them to the uterus. The physiological basis for this differential retention has yet to be clarified. Degenerating eggs can be flushed from the uterine lumen as late as 15 days after ovulation in sheep (Bindon 1969), but such eggs are probably expelled from the uterus before the next period of oestrus. The presence of a zona pellucida may protect the degenerating cytoplasm from phagocytosis.

REFERENCES

Adams CE, Chang MC (1962) Capacitation of rabbit spermatozoa in the Fallopian tube and in the uterus. J Exp Zool 15:159–166
Ainsworth L, Baker RD, Armstrong DT (1975) Pre-ovulatory changes in follicular fluid prostaglandin F levels in swine. Prostaglandins 9:915–925
Aitken RJ, Kelly RW (1985) Analysis of the direct effects of prostaglandins on human sperm function. J Reprod Fertil 73:139–146
Austin CR (1951) Observations on the penetration of the sperm into the mammalian egg. Aust J Sci Res B 4:581–596
Austin CR (1952) The "capacitation" of the mammalian sperm. Nature (Lond) 170:326
Austin CR (1956) Cortical granules in hamster eggs. Exp Cell Res 10:533–540
Austin CR (1959) Entry of spermatozoa into the Fallopian tube mucosa. Nature (Lond) 183:908–909
Austin CR (1961) The mammalian egg. Blackwell Sci Publ Oxford
Austin CR (1963a) Fertilization and transport of the ovum. In: Hartman, CG (ed) Mechanisms concerned with conception. Pergamon Press, Oxford, pp 285–320
Austin CR (1963b) Acrosome loss from the rabbit spermatozoon in relation to entry into the egg. J Reprod Fertil 6:313–314
Austin CR (1964) Behaviour of spermatozoa in the female genital tract and in fertilization. Proc 5th Int Congr Anim Reprod, Trento 3:7–22
Austin CR (1965) Fertilization. Prentice-Hall, New Jersey, USA
Austin CR (1967) Capacitation of spermatozoa. Int J Fertil 12:25–31
Austin CR (1975) Membrane fusion events in fertilization. J Reprod Fertil 44:155–166
Austin CR, Bavister BD (1974) Preliminaries to the acrosome reaction in mammalian spermatozoa. In: Afzelius B (ed) The functional anatomy of the spermatozoon. Wenner-Green Institute, Stockholm
Austin CR, Braden AWH (1953) An investigation of polyspermy in the rat and rabbit. Austral J Biol Sci 6:674–692

Austin CR, Braden AWH (1954) Time relations and their significance in the ovulation and penetration of eggs in rats and rabbits. Austral J Biol Sci 7:179–194

Austin CR, Walton A (1960) Fertilization. In: Parkes AS (ed) Marshall's physiology of reproduction. Longmans Green, London, vol 1, part 2, pp 310–416

Barros C, Yanagimachi R (1971) Induction of the zona reaction in golden hamster eggs by cortical granule material. Nature (Lond) 233:268–269

Barros C, Bedford JM, Franklin LE, Austin CR (1967) Membrane vesiculation as a feature of the mammalian acrosome reaction. J Cell Biol 34:C1–C5

Bavister BD (1969) Environmental factors important for in vitro fertilization in the hamster. J Reprod Fertil 18:544–545

Bavister BD, Teichman RJ (1976) Capacitation of hamster spermatozoa with adrenal gland extracts. Biol Reprod 14:219–221

Bavister BD, Chen AF, Fu PC (1979) Catecholamine requirement for hamster sperm motility in vitro. J Reprod Fertil 56:507–513

Beatty RA (1957) Parthenogenesis and polyploidy in mammalian development. Cambridge University Press

Beatty RA (1961) Genetics of mammalian gametes. Anim Breed Abstr 29:243–256

Bedford JM (1967) Experimental requirement for capacitation and observations on ultra-structural changes in rabbit spermatozoa during fertilization. J Reprod Fertil Suppl 2:35–48

Bedford JM (1968a) Importance of the Fallopian tube for capacitation in the rabbit. Proc 6th Int Congr Anim Reprod, Paris 1:35–37

Bedford JM (1968b) Ultrastructural changes in the sperm head during fertilisation in the rabbit. Am J Anat 123:329–358

Bedford JM (1970) Sperm capacitation and fertilization in mammals. Biol Reprod Suppl 2:128–158

Bedford JM (1972) An electron-microscopic study of sperm penetration into the rabbit egg after natural mating. Am J Anat 133:213–254

Bedford JM (1982) Fertilization. In: Austin CR, Short RV (eds) Reproduction in mammals. Cambridge University Press, book 1, pp 128–163

Bedford JM (1983) Significance of the need for sperm capacitation before fertilization in eutherian mammals. Biol Reprod 28:108–120

Betteridge KJ (1977) Embryo transfer in farm animals. Can Dept Agric Monogr No 16

Betteridge KJ, Mitchell D (1974) Direct evidence of retention of unfertilized ova in the oviduct of the mare. J Reprod Fertil 39:145–148

Betteridge KJ, Raeside JI (1962) Observation of the ovary by peritoneal cannulation in pigs. Res Vet Sci 3:390–398

Bindon BM (1969) Fate of the unfertilized sheep ovum. J Reprod Fertil 20:183–184

Bishop DW, Tyler A (1956) Fertilizins of mammalian eggs. J Exp Zool 132:575–601

Bishop MWH (1964) Paternal contribution to embryonic death. J Reprod Fertil 7:383–396

Blandau RJ (1955) Ovulation in the living albino rat. Fertil Steril 6:391–404

Bomsel-Helmreich O (1965) Heteroploidy and embryonic death. In: Wolstenholme GEW, O'Connor M (eds) Preimplantation stages of pregnancy. Churchill, London, pp 246–267 (Ciba Foundation Symposium)

Bomsel-Helmreich O (1971) Fate of heteroploid embryos. In: Raspé G (ed) Schering Symposium on Intrinsic and extrinsic factors in early mammalian development. Adv Biosci 6:381–401

Bomsel-Helmreich O (1985) Ultrasound and the preovulatory human follicle. Oxford Rev Reprod Biol 7:1–72

Braden AWH, Austin CR (1954) The number of sperms about the eggs in mammals and its significance for normal fertilisation. Austral J Biol Sci 7:543–551

Braden AWH, Austin CR, David HA (1954) The reaction of the zona pellucida to sperm penetration. Austral J Biol Sci 7:391–409

Brown CR, Harrison RAP (1978) The activation of proacrosin in spermatozoa from ram, bull and boar. Biochim Biophys Acta 526:202–217

Calvin HI, Bedford JM (1971) Formation of disulphide bonds in the nucleus and accessory structures of mammalian spermatozoa during maturation in the epididymis. J Reprod Fertil Suppl 13:65–75

Chang MC (1951) Fertilizing capacity of spermatozoa deposited into the Fallopian tubes. Nature (Lond) 168:697–698

Chang MC (1957) A detrimental effect of seminal plasma on the fertilizing capacity of sperm. Nature (Lond) 179:258–259

Chang MC, Hunter RHF (1975) Capacitation of mammalian sperm: biological and experimental aspects. In: Hamilton DW, Greep RO (eds) Handbook of physiology, Endocrinology V. American Physiological Society, Washington, ch 16, pp 339–351

Chang MC, Pincus G (1951) Physiology of fertilization in mammals. Physiol Rev 31:1–26

Chang MC, Austin CR, Bedford JM, Brackett BG, Hunter RHF, Yanagimachi R (1977) Capacitation of spermatozoa and fertilisation in mammals. In: Greep RO, Koblinsky MA (eds) Frontiers in reproduction and fertility control. MIT, Cambridge, pp 435–451

Channing CP (1966) Progesterone biosynthesis by equine granulosa cells growing in tissue culture. Nature (Lond) 210:1266

Cohen J, Adeghe AJH (1987) The other spermatozoa: fate and functions. In: Mohri H (ed) New horizons in sperm cell research. Jpn Sci Soc pp 125–134

Crozet N, Dumont M (1984) The site of the acrosome reaction during in vivo penetration of the sheep oocyte. Gamete Res 10:97–105

Cummins JM, Yanagimachi R (1982) Sperm-egg ratios and the site of the acrosome reaction during in vivo fertilization in the hamster. Gamete Res 5:239–256

Demoulin A (1985) Is diagnostic ultrasound safe during the periovulatory period? Res Reprod 17:1–2

Dickmann Z (1962) Experiments on interspecific sperm penetration through the zona pellucida. J Reprod Fertil 4:121–124

Dickmann Z (1964) The passage of spermatozoa through and into the zona pellucida of the rabbit egg. J Exp Biol 41:177–182

Dickmann Z (1965) Sperm penetration into and through the zona pellucida of the mammalian egg. In: Wolstenholme GEW, O'Connor M (eds) Preimplantation stages of pregnancy. Churchill, London, pp 169–178 (Ciba Foundation Symposium)

Dunbar BS (1983) Morphological, biochemical and immunochemical characterization of the mammalian zona pellucida. In: Hartmann JF (ed) Mechanism and control of animal fertilization. Academic Press, New York, pp 139–175

Dziuk PJ (1960) Frequency of spontaneous fragmentation of ova in unbred gilts. Proc Soc Exp Biol Med 103:91–92

Dziuk PJ (1965) Time of maturation and fertilisation of the sheep egg. Anat Rec 153:211–224

Dziuk PJ (1970) Estimation of optimum time for insemination of gilts and ewes by double-mating at certain times relative to ovulation. J Reprod Fertil 22:277–282

Fechheimer NS, Beatty RA (1974) Chromosomal abnormalities and sex ratio in rabbit blastocysts. J Reprod Fertil 37:331–341

Fléchon JE (1970) Nature glycoprotéique des granules corticaux de l'oeuf de lapine. J Microscop 9:221–242

Fléchon JE (1985) Sperm surface changes during the acrosome reaction as observed by freeze-fracture. Am J Anat 174:239–248

Fraser LR (1984) Mechanisms controlling mammalian fertilization. Oxford Rev Reprod Biol 6:174–225

Gates AH (1965) Rate of ovular development as a factor in embryonic survival. In: Wolstenholme GEW, M O'Connor (eds) Preimplantation stages of pregnancy. Churchill, London, pp 270–293 (Ciba Foundation Symposium)

Gulyas BJ (1980) Cortical granules of mammalian eggs. Int Rev Cytol 65:357–392

Guraya SS (1983) Recent progress in the structure, origin, composition and function of cortical granules in animal eggs. Int Rev Cytol 78:257–360

Gwatkin RBL (1977) Fertilisation mechanisms in man and mammals. Plenum Press, New York

Gwatkin RBL, Williams DT (1974) Heat sensitivity of the cortical granule protease from hamster eggs. J Reprod Fertil 39:153–155

Gwatkin RBL, Williams DT, Hartmann JF, Kniazuk M (1973) The zona reaction of hamster and mouse eggs: production in vitro by a trypsin-like protease from cortical granules. J Reprod Fertil 32:259–265

Hancock JL (1959) Polyspermy of pig ova. Anim Prod 1:103–106

105

Hancock JL (1961) Fertilization in the pig. J Reprod Fertil 2:307−331

Hancock JL (1962) Fertilization in farm animals. Anim Breed Abstr 30:285−310

Harper MJK (1961) The mechanisms involved in the movement of newly ovulated eggs through the ampulla of the rabbit Fallopian tube. J Reprod Fertil 2:522−524

Harrison RAP (1977) The metabolism of mammalian spermatozoa. In: Greep RO, Koblinsky MA (eds) Frontiers in reproduction and fertility control. MIT, Cambridge, pp 379−401

Harrison RAP, Fléchon JE, Brown CR (1982) The location of acrosin and proacrosin in ram spermatozoa. J Reprod Fertil 66:349−358

Hartmann JF (1983) Mammalian fertilization: gamete surface interactions in vitro. In: Hartmann JF (ed) Mechanism and control of animal fertilization. Academic Press, New York, pp 325−364

Hartree EF (1977) Spermatozoa, eggs and proteinases. Biochem Soc Trans 5:375−394

Herz Z, Northey D, Lawyer M, First NL (1985) Acrosome reaction of bovine spermatozoa in vivo: sites and effects of stages of estrous cycle. Biol Reprod 32:1163−1168

Hunter RHF (1967a) The effects of delayed insemination on fertilization and early cleavage in the pig. J Reprod Fertil 13:133−147

Hunter RHF (1967b) Effect of ageing eggs on embryonic survival. J Anim Sci 26:945

Hunter RHF (1969) Capacitation in the golden hamster with special reference to the influence of the uterine environment. J Reprod Fertil 20:223−237

Hunter RHF (1972) Fertilisation in the pig: sequence of nuclear and cytoplasmic events. J Reprod Fertil 29:395−406

Hunter RHF (1973) Polyspermic fertilization in pigs after tubal deposition of excessive numbers of spermatozoa. J Exp Zool 183:57−64

Hunter RHF (1978) Intraperitoneal insemination, sperm transport and capacitation in the pig. Anim Reprod Sci 1:167−179

Hunter RHF (1979) Fertilisation studies: a review. Anim Reprod Sci 1:251−253

Hunter RHF (1980) Physiology and technology of reproduction in female domestic animals. Academic Press, London

Hunter RHF (1984) Pre-ovulatory arrest and peri-ovulatory redistribution of competent spermatozoa in the isthmus of the pig oviduct. J Reprod Fertil 72:203−211

Hunter RHF (1986) Peri-ovulatory physiology of the oviduct with special reference to sperm transport, storage and capacitation. Dev Growth & Differ Suppl 28:5−7

Hunter RHF (1987a) Peri-ovulatory physiology of the oviduct, with special reference to progression, storage, and capacitation of spermatozoa. In: Mohri H (ed) New horizons in sperm cell research. Jpn Sci Soc pp 31−45

Hunter RHF (1987b) The timing of capacitation in mammalian spermatozoa − a reinterpretation. Res Reprod 19:3−4

Hunter RHF, Hall JP (1974) Capacitation of boar spermatozoa: synergism between uterine and tubal environments. J Exp Zool 188:203−214

Hunter RHF, Nichol R (1983) Transport of spermatozoa in the sheep oviduct: preovulatory sequestering of cells in the caudal isthmus. J Exp Zool 228:121−128

Hunter RHF, Poyser NL (1985) Ovarian follicular fluid concentrations of prostaglandins E_2, $F_{2\alpha}$ and I_2 during the pre-ovulatory period in pigs. Reprod Nutr Dev 25:909−917

Hunter RHF, Holtz W, Herrmann H (1978) Stabilising rôle of epididymal plasma in relation to the capacitation time of boar spermatozoa. Anim Reprod Sci 1:161−166

Ito S, Kudo A, Niwa T (1959) Studies on the normal oestrus in swine with special reference to proper time for service. Annls Zootech Ser D, Suppl pp 105−107

Katz DF, Yanagimachi R, Dresdner RD (1978) Movement characteristics and power output of guinea-pig and hamster spermatozoa in relation to activation. J Reprod Fertil 52:167−172

Kelly RW (1981) Prostaglandin synthesis in the male and female reproductive tract. J Reprod Fertil 62:293−304

Leese HJ, Barton AM (1985) Production of pyruvate by isolated mouse cumulus cells. J Exp Zool 234:231−236

LeMaire WJ, Yang NST, Behrman HR, Marsh JM (1973) Preovulatory changes in the concentration of prostaglandins in rabbit Graafian follicles. Prostaglandins 3:367−376

Lillie FR (1912) The production of sperm iso-agglutinins by ova. Science 36:527−530

Lovell JE, Getty R (1968) Fate of semen in the uterus of the sow: histologic study of endometrium during the 27 hours after natural service. Am J Vet Res 29:609−625

Meizel S (1984) The importance of hydrolytic enzymes to an exocytotic event, the mammalian sperm acrosome reaction. Biol Rev 59:125−157

Meizel S, Lui CW, Working PK, Mrsny RJ (1980) Taurine and hypotaurine: their effects on motility, capacitation and acrosome reaction of hamster sperm in vitro and their presence in sperm and reproductive tract fluids of several mammals. Dev Growth & Diff 22:283−294

Miyazaki S, Igusa Y (1981) Fertilization potential in golden hamster eggs consists of recurring hyperpolarizations. Nature (Lond) 290:702−704

Mortimer D (1978) Selectivity of sperm transport in the female genital tract. In: Cohen J, Hendry WF (eds) Spermatozoa, antibodies and infertility. Blackwells, Oxford, pp 37−53

Nicolson GL, Yanagimachi R (1972) Terminal saccharides on plasma membranes: identification by specific agglutinins. Science 177:276−279

Nicolson GL, Usui N, Yanagimachi R, Yanagimachi H, Smith JR (1977) Lectin-binding sites on the plasma membranes of rabbit spermatozoa. J Cell Biol 74:950−962

Oprescu St, Thibault C (1965) Duplication de l'ADN dans les oeufs de lapine après la fécondation. Annls Biol Anim Biochim Biophys 5:151−156

Overstreet JW, Bedford JM (1975) The penetrability of rabbit ova treated with enzymes or anti-progesterone antibody: a probe into the nature of a mammalian fertilizin. J Reprod Fertil 44:273−284

Piko L (1969) Gamete structure and sperm entry in mammals. In: Metz CB, Monroy A (eds) Fertilization: comparative morphology, biochemistry and immunology. Academic Press, New York 2:325−403

Piko L, Tyler A (1964) Fine structural studies of sperm penetration into the rat. Proc 5th Int Congr Anim Reprod, Trento 2:372−377

Pitkjanen IG (1958) The time of ovulation in sows. Svinovodstvo 12:38−40

Polge C, Salamon S, Wilmut I (1970) Fertilizing capacity of frozen boar semen following surgical insemination. Vet Rec 87:424−428

Rasweiler JJ (1972) Reproduction in the long-tongued bat, *Glossophaga soricina*. 1. Preimplantation development and histology of the oviduct. J Reprod Fertil 31:249−262

Rogers BJ (1978) Mammalian sperm capacitation and fertilisation in vitro: a critique of methodology. Gamete Res 1:165−223

Rowson LE (1951) Methods of inducing multiple ovulation in cattle. J Endocrinol 7:260−270

Sacco AG, Yurewicz EC, Subramanian MG (1986) Carbohydrate influences the immunogenic and antigenic characteristics of the ZP3 macromolecule of the pig zona pellucida. J Reprod Fertil 76:575−586

Stambaugh R, Noriega C, Mastroianni L (1969) Bicarbonate ion: the corona cell dispersing factor of rabbit tubal fluid. J Reprod Fertil 18:51−58

Suarez SS (1987) Sperm transport and motility in the mouse oviduct: observations in situ. Biol Reprod 36:203−210

Szollosi D (1962) Cortical granules: a general feature of mammalian eggs? J Reprod Fertil 4:223−224

Szollosi D (1966) Time and duration of DNA synthesis in rabbit eggs after sperm penetration. Anat Rec 154:209−212

Szollosi D (1967) Development of cortical granules and the cortical reaction in rat and hamster eggs. Anat Rec 159:431−446

Szollosi D (1975) Mammalian eggs ageing in the Fallopian tubes. In: Blandau RJ (ed) Ageing Gametes. Karger, Basel, pp 98−121

Szollosi D, Hunter RHF (1973) Ultrastructural aspects of fertilization in the domestic pig: sperm penetration and pronucleus formation. J Anat 116:181−206

Szollosi D, Hunter RHF (1978) The nature and occurrence of the acrosome reaction in spermatozoa of the domestic pig, *Sus scrofa*. J Anat 127:33−41

Szollosi DG, Ris H (1961) Observations on sperm penetration in the rat. J Biophys Biochem Cytol 10:275−283

Templeton AA, Mortimer D (1980) Laparoscopic sperm recovery in infertile women. Brit J Obstet Gynaecol 87:1128−1131

Thibault C (1959) Analyse de la fécondation de l'oeuf de la truie aprés accouplement ou insemination artificelle. Annls Zootech Ser D, Suppl pp 165–177

Thibault C (1967) Analyse comparée de la fécondation et de ses anomalies chez la brebis, la vache et la lapine. Annls Biol Anim Biochim Biophys 7:5–23

Thibault C (1972) Some pathological aspects of ovum maturation and gamete transport in mammals and man. Acta Endocrinol Suppl 166:59–66

Thibault C, Dauzier L (1960) "Fertilisines" et fécondation in vitro de l'oeuf de lapine. CR Acad Sci, Paris 250:1358–1359

Tsafriri A, Lindner HR, Zor U, Lamprecht SA (1972) Physiological rôle of prostaglandins in the induction of ovulation. Prostaglandins 2:1–10

Van Niekerk CH, Gerneke WH (1966) Persistence and parthenogenetic cleavage of tubal ova in the mare. Onderstepoort J Vet Res 33:195–231

Yanagimachi R (1969a) In vitro capacitation of hamster spermatozoa by follicular fluid. J Reprod Fertil 18:275–286

Yanagimachi R (1969b) In vitro acrosome reaction and capacitation of golden hamster spermatozoa by bovine follicular fluid and its fractions. J Exp Zool 170:269–280

Yanagimachi R (1970) The movement of golden hamster spermatozoa before and after capacitation. J Reprod Fertil 23:193–196

Yanagimachi R (1981) Mechanisms of fertilization in mammals. In: Mastroianni L, Biggers JD (eds) Fertilization and embryonic development in vitro. Plenum, New York, pp 81–182

Yanagimachi R, Noda YD (1970a) Ultrastructural changes in the hamster sperm head during fertilization. J Ultrastruct Res 31:465–485

Yanagimachi R, Noda YD (1970b) Electron microscopic studies of sperm incorporation into the golden hamster egg. Am J Anat 128:429–462

Yanagimachi R, Usui N (1974) Calcium dependence of the acrosome reaction and activation of guinea-pig spermatozoa. Expl Cell Res 89:161–174

CHAPTER VI

Development of the Embryo and Influences on the Maternal System

CONTENTS

Introduction 109
Developmental Stages in the Fallopian Tube 110
Nutritional Aspects of Embryonic Development 112
Early Pregnancy Factors 115
Experimental Modification of Tubal Development 120
Organization of the Embryo: Some First Steps 120
References 122

Introduction

That the normally fertilized egg undergoes two or more cleavage divisions (mitotic divisions) in the Fallopian tube before passing to the uterus has been widely recognized for many years (Table V.2); indeed, classical nineteenth century studies recorded this fact, which was to be endorsed and given considerable detail in the early part of the twentieth century (see Austin 1961). Perhaps a high point of this phase of morphological study was the publication of a monograph entitled *The Eggs of Mammals* (Pincus 1936), with much of the information being subsequently reviewed by Chang and Pincus (1951). Professor G. W. Corner, in whose Rochester laboratory the steroid hormone progesterone was first isolated, was also well aware of the cleavage divisions of the fertilized egg from extensive studies in domestic animals and primates. But, rather than interpreting such development as a carefully orchestrated event occurring within a specialized fluid environment, his view was that the embryo − at least in one sense − was biding its time until the glandular epithelium of the uterus and its secretions had developed under the influence of ovarian progesterone; uterine secretions were needed in order to stimulate and sustain the cleaving embryo (Corner 1942). Since the late 1960s, however, it has been appreciated that the mammalian embryo is internally programmed by its genome to follow a strict chronological sequence of development and externally programmed by the availability of substrates in the luminal fluids of the Fallopian tube. Detailed information on different substrates and the patterns of metabolism in pre-implantation embryos has been derived from culture studies in chemically-defined media, initially using large numbers of rodent embryos (Brinster 1965, 1973), then with smaller numbers of embryos from species such as sheep (Tervit et al. 1972; Tervit and Rowson 1974; Wright and Bondioli 1981), and more recently still with human embryos obtained by means of in vitro fertilization (Edwards and Purdy 1982; Beier and Lindner 1983; Wood and Trounson 1984; Testart and Frydman 1985; Leese et al. 1986).

This chapter attempts to summarize some significant points in development of the mammalian embryo, and especially to consider possible influences of the cleaving embryo upon maternal physiology. Appropriate reviews of the literature are referred to in the pages that follow; a useful starting point would be McLaren (1982).

Developmental Stages in the Fallopian Tube

The most advanced stage of development achieved during passage of embryos along the Fallopian tubes seems not to be directly related to the gestation length of the species under consideration, but it may reflect the rate of passage to the uterus. In laboratory species such as rats, mice, guinea-pigs and rabbits, embryos remain for approximately 3 days in the tubes and enter the uterus as young morulae of 8 – 16 cells. Similarly, the embryos of ruminants such as the sheep and cow enter the uterus approximately 3 days after ovulation at the stage of 8 – 16 cells. By contrast, those of the pig take only 2 days and are still at the 4-cell stage when leaving the tubes (Assheton 1898; Pomeroy 1955). Estimates for the rate of embryo passage in the horse suggest some 6 days, with the uterine lumen being reached at the late morula or blastocyst stage (Oguri and Tsutsumi 1972; Betteridge, Eaglesome and Flood 1979). The embryos of certain carnivores, such as ferrets and mink, may spend equally long or even longer in the Fallopian tube, perhaps remaining there for some 5 – 6 days and 8 days, respectively, until they have developed into blastocysts (Hansson 1947; Chang 1969, 1973).

As to the rate of egg passage in primates, figures currently available suggest that a 3 – 4-day sojourn in the Fallopian tubes is most common in rhesus monkeys and women, with morulae of 8 – 16 cells passing to the uterus (Harper and Pauerstein 1976; Edwards 1980; Biggers 1981; Hearn 1986); these estimates agree with the earlier values of Hertig, Rock and Adams (1956). It is worth remarking here that although the time of ovulation in women can now be gauged with accuracy by means of real-time echo sonography of the ovarian structures (see Chap. IV), there is as yet no comparable means of judging precisely when the embryo has passed from the Fallopian tube to the uterus.

Fertilized and unfertilized eggs of nearly all placental mammals so far examined have been thought to proceed to the uterus on a similar time scale. This is not the case, however, in the horse and certain other equids in which unfertilized eggs remain in the Fallopian tube for a protracted period of time. Various species of bat may also exhibit this distinction (see Chap. V). Recent work in hamsters has noted differences between fertilized and unfertilized eggs in their rate of transport to the uterus (Ortiz et al. 1986), leading to consideration of locally mediated signals: early pregnancy factors, as discussed below, may be of relevance here.

Conclusions *not* to be drawn from the preceding data on developmental stages are:

a) that the duration of the 2-cell, 4-cell and 8-cell stages is similar; this is not so. As is known, for example, from studies in pigs, the 2-cell stage is relatively short, lasting for only 5 – 6 hours, whereas the 4-cell stage requires 24 h or more (Hunter 1974; Polge 1982);

b) that cleavage proceeds evenly through the stages 2-cell, 4-cell, 8-cell etc; this is not always the case, and embryos composed of 3, 5 and 7 blastomeres are frequently observed;

c) that all embryos recovered from the same tube in polyovular species will be at similar stages; this is seldom so, as is revealed grossly by the number of blastomeres or, using phase-contrast microscopy, by the number of nuclear structures.

The mammalian oocyte is a remarkably large cell (approximately 80 – 120 µm in diameter with the zona pellucida), but cleavage divisions of the embryo progressively reduce the size of the component blastomeres and increase the ratio of nuclear to cytoplasmic material. By the 8 – 16-cell stage, morulae are being formed in which component cells undergo a modification from spherical- to wedge-shaped, thereby achieving a more intimate contact between individual cells as flattening occurs. This process of compaction gives the cells an orientation or polarity which is considered vital for elaboration of different cell types (see Johnson 1979).

The stage of development at which differentiation commences has long been of interest, but there is sound experimental evidence from farm and laboratory animals indicating that blastomeres can express totipotency until the 2- or 4-cell stage and sometimes one cleavage division later. This has been demonstrated either by microsurgical destruction of individual blastomeres or, more elegantly, by microsurgical separation of such blastomeres and the subsequent development of the separated cells under appropriate conditions involving transplantation into recipient animals (Moore et al. 1968, 1969). In fact, this has been the means of producing identical twins, triplets or even quadruplets in farm animals (Trounson and Moore 1974; Willadsen 1979, 1981), although the extent to which the technique has been applied to human embryos − if at all − remains unknown. Totipotency as demonstrated in the above manner does not necessarily mean that differentiation has not yet commenced by the 4-cell stage, but it does indicate that the process has not reached an irreversible stage. Even so, a predetermination of cell lines can frequently be demonstrated after the 2-cell stage in mouse embryos (Johnson 1981).

With the notable exception of some species of bat (Rasweiler 1979; van der Merwe 1982), the cleaving embryo remains within the zona pellucida whilst in the Fallopian tube of nearly all eutherian mammals so far examined. This relatively thick non-cellular coating doubtless protects the embryo in diverse ways, not least in preventing contact of the true embryonic surface with other embryos or with the tubal epithelium which might lead to adhesion and thereby to interference with transport. Whether there is also a component of immunological masking of cell surface antigens remains uncertain. However, a major function of the zona pellucida appears to be the purely physical one of holding the blastomeres together until sufficient intercellular bridges have developed to anchor the cells

111

in a suitable relationship with each other. Edwards (1964) found that liberation of 2-cell rabbit embryos from the zona pellucida by pronase dissolution followed by a period of culture in vitro led to disaggregation of cleaving embryos, a situation also recorded for 8-cell sheep embryos (Moor and Cragle 1971). Mintz (1962) had similarly noted abnormal cleavage patterns in mouse embryos in vitro after pronase removal of the zona pellucida. As a related observation after microsurgery on 2- and 4-cell embryos for the purpose of producing identical offspring in farm animals, the resultant individual blastomeres required to be installed in 'host' zonae sealed with agar in order for embryos to proceed successfully through the early cleavage stages (Willadsen 1979).

A further function of the zona pellucida, all too often overlooked and yet especially relevant to human fertility, is to protect the embryo from subsequent penetration by spermatozoa. Exposure to viable spermatozoa may continue long after the initial stages of fertilization, either by contact of the embryo with sperm cells migrating from reservoirs in the caudal portion of the isthmus (Chap. IV) or because of a subsequent bout of coitus. It may be inferred, therefore, that the block to polyspermy has a protracted stability, and this has been demonstrated by surgical insemination directly into the Fallopian tubes of pigs when embryos were already at the 2- and 4-cell stages (unpublished observations).

In normal circumstances, escape of the embryo from the zona pellucida — sometimes referred to as hatching — is a uterine event which usually occurs at the blastocyst stage of development and precedes implantation or attachment. Indeed, significant growth of the embryo can take place only after escape from the zona pellucida, this being particularly true in a number of species of ungulate (e.g. cow, sheep, pig) in which dramatic elongation of the embryo into a structure termed a conceptus is found prior to attachment.

Evolution from the spherical to the elongated or filamentous form is associated with a specific cellular remodelling in the endoderm and trophectoderm. The surface of the zona pellucida may have become coated with a layer of mucin during passage along the Fallopian tube, as in the rabbit (Adams 1958; Greenwald 1958), so the rôle of this layer both in the tube and at the time of hatching needs also to be considered.

Nutritional Aspects of Embryonic Development

As already noted, mammalian oocytes are large cells and the cytoplasm of the freshly-ovulated egg contains substrates for use in metabolic activities: many of these accumulate during growth of the primary oocyte. Other constituents, and especially a broad spectrum of proteins, become apparent after the pre-ovulatory gonadotrophin surge and shortly before ovulation (Warnes, Moor and Johnson 1977; Moor and Warnes 1978). In the eggs of farm animals, and most conspicuously in those of the pig, there are considerable reserves of lipids — sometimes referred to as yolk or deutoplasm — and these give the cytoplasm a dense and relatively dark appearance. But, despite the presence of nutritional reserves in the newly-shed mammalian egg, these unique cells may depend critically upon energy substrates in the fluids of the Fallopian tube, this being so even before the occurrence of sperm penetration and egg activation. In several species, the requirements of oocytes and embryos in vivo have been deduced from culture studies using relatively simple chemically defined media. However, in other species, progression through the early cleavage stages has been possible in vitro only in the presence of tissue dissected from the Fallopian tube (reviews by Whitten 1971; Biggers and Borland 1976).

Due principally to detailed studies of mouse eggs in vitro, it is now known that maturation of pre-ovulatory oocytes can take place in a basic salt solution with pyruvate as the sole energy source (Brinster 1965). If glucose is substituted for pyruvate as the energy substrate, maturation of oocytes in vitro will not occur in the absence of surrounding follicular cells which elaborate pyruvate from the glucose and transmit it to the oocytes (Donahue and Stern 1968; Leese and Barton 1985). Energy metabolism of unfertilized eggs is thus seen to be highly specialized, and parallel evidence has been produced for the zygote, suggesting a crucial rôle for the Fallopian tube fluids in the nutritional support of the developing embryo. As again observed with in vitro studies on mice (Biggers, Whittingham and Donahue 1967; Brinster 1973), the zygote can divide only when pyruvate or oxaloacetate are present in the culture medium whereas phosphoenolpyruvate and lactate can also be used as substrates after the 2-cell stage has been reached (Fig. VI.1); glucose cannot be used until the 8-cell stage. The concentration of pyruvate in tubal fluid increases considerably after ovulation (see comment below concerning follicular cells) but, significantly, changes in lactate concentration are the most striking. These are limited examples, the essential point being that the nature of the substrates available in tubal fluid is regulated by the prevailing balance of ovarian hormones, and is therefore changing in step with the embryos' requirement for and ability to use the substrates (see Hunter 1977). This observation is supported by the results of egg and embryo transplantation studies in laboratory and farm animals: optimum survival of the transplanted egg requires a close synchrony in the stage of the oestrous cycle between donor and recipient (Chang 1950), inferring critical changes in the fluid medium of the free-living embryo.

The potential rôle of the granulosa cells liberated from the egg surface should not be overlooked in a discussion of nutrition of the embryo (see Biggers et al. 1967). Because these somatic cells (1) are involved in growth and metabolism of the oocyte before ovulation, (2) can be cultured in vitro and show synthetic activity, and (3) in some species, at least, remain in the Fallopian tubes in the vicinity of the embryos (Hunter 1978), their contribution to the fluid microenvironment should certainly be considered. The ability of granulosa cells to synthesize and secrete steroids from appropriate substrates could modify the epithelium of the tube, and their potential for converting glucose into pyruvate may be significant (Leese and Barton 1985). In any event, it is difficult to imagine that the fate of the liberated follicular granulosa cells is signalled at the time of ovulation.

The content of bicarbonate, amino acids and diverse minerals (cations) in tubal fluid has now been examined in some detail (see Chap. III), but a specific involvement of such constituents in individual cleavage stages is less well documented. Culture media usually contain bicarbonate − in conjunction with a CO_2 gas phase − for regulation of pH, and blood serum or serum albumin are invariably present as a protein source. Fluids for culture are classical tissue culture

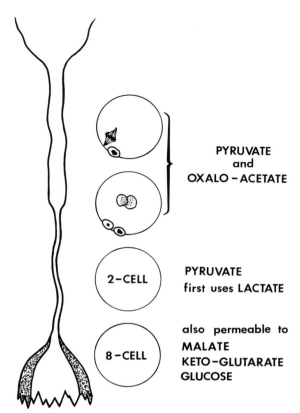

Fig. VI.1. Diagrammatic representation of a Fallopian tube to indicate the different metabolic substrates used by the unfertilized egg and cleaving rodent zygote. The more advanced the stage of development, the greater the range of nutritional substrates incorporated by the embryo

**PYRUVATE
and
OXALO – ACETATE**

2 – CELL

**PYRUVATE
first uses LACTATE**

8 – CELL

**also permeable to
MALATE
KETO – GLUTARATE
GLUCOSE**

media, such as TC 199, Ham's F 10 or Earle's solution with appropriate supplementation, but a medium has been formulated based upon actual analysis of Fallopian tube fluid (Menezo and Laviolette 1972).

A local influence of the embryo on the synthetic activity of the tubal epithelium was suggested by Stone and Hamner (1975) and Hunter (1977), but the evidence remains equivocal. The report by Roblero, Biggers and Lechene (1976) that the ionic compositions of ampullary and isthmic fluid are dissimilar, and that both are very different in composition from blood plasma, could be interpreted in support of such local effects (Chap. III). Likewise, differences in composition of the microenvironment surrounding cleaving mouse embryos and that in other portions of the tubal lumen suggest a local influence of the embryo and/or the liberated cumulus cells (Borland et al. 1977). Nonetheless, the critical question remains as to when the pre-implantation embryo first dictates or modifies, (if at all), the pattern of nutritional substrates available from the mother. Use of increasingly-sophisticated technology to examine microenvironments in the vicinity of the embryo may reveal an embryonic influence shortly after completion of fertilization.

The energy requirements of human oocytes and embryos in vitro have yet to be documented in detail, although published results on the culture of embryos are now accumulating (Edwards 1980; Fishel, Edwards and Purdy 1983; Mohr and Trounson 1984; Leese et al. 1986). Adequate numbers of eggs will be required before appropriate conclusions can be drawn, and yet obtaining such material and undertaking the culture studies raise considerable problems, not least those of an ethical nature. In any case, up to the time of writing, culture studies appear to have been more concerned with promoting normal cleavage of embryos in quite elaborate culture media, such as modified Whitten's or Ham's F10, rather than in determining optimum energy sources or amino acid requirements. One of the earliest reports was that of Edwards, Steptoe and Purdy (1970) documenting in vitro cleavage of in vitro fertilized human eggs up to the blastocyst stage, using Ham's F10 medium supplemented with foetal calf and human serum. However, it certainly remains questionable whether there are specific secretions of the Fallopian tube necessary for development of the primate embryo. Transfer of human oocytes and spermatozoa directly into the uterus followed by successful establishment of pregnancy (Craft et al. 1982) infers, at the very least, a considerable degree of interchangeability between tubal and uterine secretions although, of course, essential components of tubal secretions might have entered the uterus from the intramural portion. With the same qualification, earlier studies by Marston, Penn and Sivelle (1977) involving autotransplantation of pronucleate eggs and young zygotes of the rhesus monkey into the uterus suggested flexibility in the fluid environment of the primate embryo.

Early Pregnancy Factors

That the embryo might influence the maternal endocrinological and/or immunological systems whilst still in the Fallopian tube has long seemed possible, and indeed diagrams of the kind portrayed in Fig. VI.2 have been used for teaching purposes in Edinburgh since the early 1970s. Under normal circumstances, the embryo contains a full diploid complement of genetic information and thus the point at issue is just how soon instructions in the genome are expressed for trophic purposes. It could be argued from several points of view that it would be biologically prudent for the embryo to influence the mother at an early stage of development. Consideration of the extremely intimate apposition of the zona pellucida and the tubal epithelium − perhaps best appreciated in scanning electron micrographs (Fig. VI.3) − also suggests the possibility of a mutual exchange of signals. Even so, one of the earliest recorded responses to a developing embryo has, until recently, been the change in permeability of the capillary bed of the uterus that precedes implantation in rodents: this can be demonstrated by the Pontamine blue reaction, a series of bands of dye in the uterine wall corresponding to the location of individual embryos, and is detectable in rats on the fourth day after mating (Psychoyos 1960, 1973).

In the domestic farm animals, a trophic influence of the embryo(s) upon ovarian function has not been reported during the 'tubal' stages, nor indeed have differences in blood flow to the capillary bed in the myosalpinx been noted in the region of the embryos. On the other hand, expression of the embryonic genome has been documented at the earliest cleavage stages in the mouse in the form of

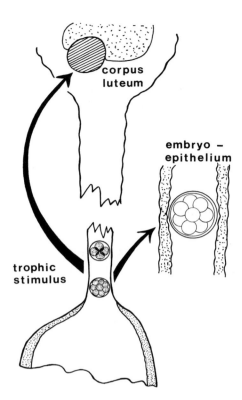

Fig. VI.2. Portrayal of portions of the Fallopian tube and corresponding ovary to suggest a possible trophic influence of products from the pre-implantation embryo on ovarian activity, especially on secretion of progesterone

a paternally-derived protein (Sawicki, Magnuson and Epstein 1981) and in other distinct control mechanisms imposed by the paternal genome (Szöllösi and Yotsuyanagi 1985). Thus, the extent to which synthetic products might influence maternal tissues in the large domestic species and primates is certainly worth considering. Pregnancy-associated proteins can be detected in the maternal circulation of mice, but this is not until Day 7 of gestation, with maximum levels appearing between Days 12 and 14 (Waites and Bell 1984). In rabbits, there have been reports that embryos may elaborate a chorionic gonadotrophin-like substance by the blastocyst stage of development (Haour and Saxena 1974; Asch et al. 1979) or indeed by the morula stage (Asch et al. 1978), in which case a trophic influence upon luteal function would be expected. Endorsement and extension of these observations by other laboratories have so far not been forthcoming, so the evidence must remain tentative.

Possible immunosuppressive effects arising from the young embryo in the Fallopian tube are thought to have been revealed by a rosette inhibition test. On the basis of such a test, alterations to the maternal lymphocytes in mice are found before the time of implantation and may be occurring within 6 h of fertilization (Morton, Hegh and Clunie 1976), a response presumed to be due to the influence of developing embryos. However, the nature and biological significance of the putative changes remain ill-defined (Whyte and Heap 1983), and there is also concern over the lack of specificity in the assay. A low molecular weight factor releas-

116

Fig. VI.3. A scanning electron micrograph of a portion of the pig Fallopian tube to show the extremely intimate contact between the epithelium of the isthmus and the developing embryo, still encompassed by the zona pellucida. (Adapted from Hunter, Fléchon and Fléchon 1987)

ed from the embryo and termed zygotin is thought to be a peptide and one component of early pregnancy factor (Nancarrow, Wallace and Grewal 1981). It can be detected in maternal blood by its synergistic action with anti-lymphocyte serum in reducing rosette formation in a rosette inhibition test, and may be revealed by the first day of embryonic life in sheep and by the fourth day in cattle (Nancarrow et al. 1981). Zygotin was suggested to act directly on the ovary or via an intermediary substance released by the Fallopian tube (Fig. VI.4). Ovum factor is another term applied to this or a similar substance (Cavanagh et al. 1982), although it is thought to stimulate production of an ovarian component of early pregnancy factor.

Whilst investigating whether other features of maternal physiology might be modified in response to the presence of pre-implantation embryos, O'Neill (1985 a) also reported that mouse embryos stimulate maternal changes within 6 h of fertilization. The changes concern populations of platelets and are expressed as an increased vascular demand for these bodies, resulting in a significant reduction in the splenic reserves and in peripheral blood platelet concentrations; this condition can therefore be referred to as early-pregnancy-associated thrombo-

117

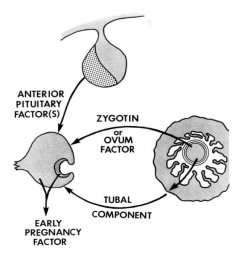

Fig. VI.4. Diagrammatic representation of the rôle of the embryo, Fallopian tube and ovary in the production of an early pregnancy factor. (Based principally on Nancarrow, Wallace and Grewal 1981)

ANTERIOR
PITUITARY
FACTOR(S)

ZYGOTIN
or
OVUM
FACTOR

TUBAL
COMPONENT

EARLY
PREGNANCY
FACTOR

cytopenia (O'Neill 1985a). Not only was this extremely early response shown to be a direct consequence of the presence of fertilized eggs, but a correlation could also be demonstrated between the degree of such thrombocytopenia and the number of embryos present in the Fallopian tube on the second day of pregnancy. This was therefore considered as an initial expression of maternal recognition of pregnancy, and of course long before the onset of implantation (O'Neill 1985a). In subsequent in vitro studies of embryos, coupled with injection of the culture medium into splenectomized mice, early-pregnancy-associated thrombocytopenia was shown to be caused by the production of platelet-activating factors by the fertilized eggs (O'Neill 1985b). Further characterization of the embryo-derived platelet-activating factor has been reported (O'Neill 1985c).

Turning the focus once more to ovarian events, inhibition of further maturation of Graafian follicles might be an expected response to fertilization, even though morphological observations of the ovaries of farm animals do not support such a proposition. But embryos of the golden hamster, a species with a short (4-day) oestrous cycle, apparently produce a tetrapeptide as early as the 2–4-cell stage of development, and this substance has been implicated in prevention of ovulation (Kent 1975).

As far as primate embryos are concerned, demonstration of one or more early pregnancy factors might form the basis of an early pregnancy test and would be invaluable in studies of early embryonic loss, either in the spontaneous situation or after transplantation procedures. Embryonic loss is frequently stated to assume a remarkably high incidence in humans (Boué, Boué and Lazar 1975; Edwards 1980; Biggers 1981; Edmonds et al. 1982; Short 1984). Using a modified rosette inhibition test, a serum factor has been claimed in women soon after fertilization, with activity being lost upon surgical abortion or apparently declining before spontaneous abortion (Morton et al. 1977; Smart et al. 1981). But it is not easy to imagine that the putative serum factor emanating from the zygote soon after fertilization would correspond precisely with a factor released by a young foetus.

118

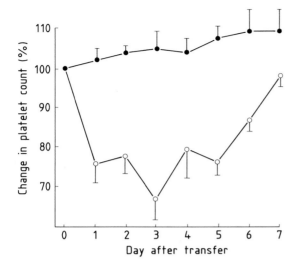

Fig. VI.5. Relative changes in the concentration of platelets in women after the transfer of embryos resulting in pregnancy (*lower curve*) or after the transfer of unfertilized eggs or culture medium (*upper curve*). (Courtesy of O'Neill et al. 1985)

As of writing, one of the earliest specific influences of the primate embryo on the maternal system that is fully accepted is that of its HCG secretion and the consequent stimulation of the corpus luteum; this occurs by the sixth day of embryonic development in women when the blastocyst is already in the uterus (review by Sauer 1979). Tentative evidence has been published for a platelet-activating factor in the peripheral blood of marmosets in very early pregnancy, and on occasions also in the culture medium used for incubating pre-implantation marmoset embryos (Hearn 1986; Hearn et al. 1986), but a more persuasive set of results is awaited. O'Neill and Saunders (1984) reported a strong correlation between a human embryo-derived platelet-activating factor and the viability of embryos — that is their potential to establish pregnancies (Fig. VI.5). Moreover, monitoring of this factor was suggested to provide a quantifiable, non-invasive means of determining which embryos are viable before transfer, and could lead subsequently to monitoring of in vivo development (O'Neill et al. 1985).

Despite the fact that much of the evidence on early pregnancy factors is less than rigorous, observations accumulating from a number of species together strongly suggest that the embryo is able to influence various components of maternal physiology whilst still in the Fallopian tube; embryonic peptide or protein molecules may provide the principal messenger system(s). Characterization of these molecules and of their maternal influences would be invaluable, and might help to explain why experimental by-passing of the Fallopian tube — by means of embryo transplantation directly into the uterus of virgin recipients — can lead to the establishment of full-term pregnancies. Although early embryonic influences within the tubal environment may be beneficial for the establishment of pregnancy, they are clearly not essential for promoting individual 'experimental' pregnancies (Chap. IX). Perhaps their contribution is best assessed in the overall breeding success of a population of mammals.

With the notable exception of the situation in certain primates (see Chap. VIII), embryos that are artificially arrested in the Fallopian tubes by means of ligatures or other forms of surgical interference seldom progress beyond the blastocyst stage; frequently, they degenerate as young morulae. This experimental situation has been examined in rabbits, rodents and farm animals (e.g. Alden 1942; Wintenberger-Torres 1956; Adams 1973); several interpretations of the findings are possible. There may be insufficient or inappropriate substrates in the decreasing volume of tubal fluid for normal development (Chap. III) or, as a related point, the embryo may have a specific requirement for uterine secretions being elaborated under the influence of increasing concentrations of plasma progesterone: in farm animals, the embryo has a rather tightly programmed requirement for interactions with the endometrium and its secretions (Miller and Moore 1976; Wilmut, Sales and Ashworth 1986). Uterine factors may be required for normal hatching from the zona pellucida (McLaren 1970), but degeneration of the embryo is visible well in advance of the hatching phenomenon.

In the light of these remarks on the limitations of the tubal environment in non-primate species, it is perhaps surprising to note that the ligated Fallopian tube of rabbits can be used as a holding site for the cleaving embryo of a number of species. Using this system of in vivo culture, cattle, sheep and pig embryos have been supported for up to 3 days, frequently during long-distance transfer experiments such as those involving intercontinental flights. Examples of the use of this technique include the pioneer studies of Averill, Adams and Rowson (1955), who reported that sheep embryos could survive for at least 5 days and develop to the early blastocyst stage in the genital tract of rabbits, and Hunter et al. (1962) who used this approach to transport 4−8-cell sheep embryos from Cambridge, UK, to Natal, South Africa. Sreenan, Scanlon and Gordon (1968) also used the pseudopregnant rabbit successfully as a culture system for cow embryos: they recorded a maximum cleavage stage of 64 blastomeres following storage of an 8-cell egg for 4 days. Lawson, Rowson and Adams (1972) have retransferred cow embryos after periods of storage in the rabbit Fallopian tube of 2 to 4 days, and have recorded the birth of a number of viable calves. As far as pig embryos are concerned, Polge, Adams and Baker (1972) have cultured the 1−4-cell stages for 2 days in rabbit tubes, and horse embryos have similarly undergone short-term culture in the rabbit, in due course yielding a viable foal upon transfer to an appropriate recipient mare (Allen et al. 1976).

Organization of the Embryo: Some First Steps

Progression through the first mitotic divisions is frequently presented − as above − as a series of cleavages involving reduction in the size of individual cells, with little hint being given as to the specialization that can be revealed so precociously. Although blastomeres may retain totipotency up to the 4-cell stage or even later, there is now persuasive evidence that differentiation − that is establishment of distinct cell lines − may already be commencing from the 2-cell stage in laborato-

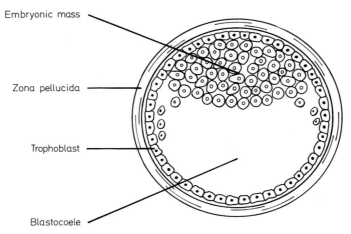

Embryonic mass

Zona pellucida

Trophoblast

Blastocoele

Fig. VI.6. Schematic portrayal of a blastocyst to illustrate the distinction between the inner cell mass or embryonic disc and cells of the trophoblast

ry rodents. Indeed, analysis of the earliest steps in formation of the embryo and the programming of cell lineage has reached a remarkably sophisticated level, largely due to the techniques of pronuclear and nuclear transplantation. The topic has been the subject of an excellent review (Surani, Barton and Norris 1987), in which the functional differences between the parental genomes received strong emphasis. In a sentence, the origin of chromosomes determines their influence during embryogenesis; the experimental evidence is cited in detail. Future analysis should be even further refined by the insertion of cloned genes rather than by the modification of haploid contributions. In the words of the above review, studies in mammals should soon progress from experimental (i.e. largely physical) to molecular embryology.

A critical functional change in the mammalian embryo after three to four cleavage divisions is a tighter bonding and stabilization of individual blastomeres by means of extensive intercellular junctions. Compaction of the embryo leads to formation of a morula, with a flattening of the outer cells and then the gradual appearance of fluid within the embryo. Fluid accumulation gives rise to the blastocoele; the surrounding cells of what is now a blastocyst have differentiated into those of the trophoblast or trophectoderm, and an inner cell mass or embryonic disc located at one pole (Fig. VI.6). The latter group of cells is destined primarily to form the embryo (and thus foetus), whereas cells of the trophectoderm give rise to embryonic membranes of the placenta. Apparently, the first cell to divide from the 2-cell mouse embryo contributes a disproportionately larger number of its progeny to the inner cell mass and fewer to the trophectoderm (McLaren 1982). The topic of early differentiation is complex, but several elegant reviews are available (Johnson 1979, 1981).

Evidence once more from the mouse indicates specific morphological changes following cleavage to the 8-cell stage, in the form of a major reorganization of the cytoplasm and cell membranes with a polarization becoming apparent in both these compartments. The microvillous attachment sites of cells become located

121

Table VI.1. Criteria for differentiating inner cell mass from trophectoderm in the $3\frac{1}{2}$-day mouse blastocyst. (After Johnson 1979)

	Inner cell mass	Trophectoderm
Structural	Focal gap junctions, no desmosomes Distinctive mitochondria Few microvilli	Cellular polarity, zonular tight junctions and desmosomes Distinctive mitochondria and cytoplasmic inclusions
Physiological	Adherent surfaces, stimulate trophectodermal proliferation, sensitive to cytotoxic drugs and high temperature	Phagocytic, transport ions and macromolecules, secrete fluid, induce decidual response
Biochemical	Distrinctive polypeptides, antigens and enzymes	Distinctive polypeptides, antigens and enzymes
Behavioural	Forms all endoderm, mesoderm and embryonic ectoderm	Forms primary and secondary trophoblastic giant cells, ectoplacental cone, extraembryonic (chorionic) ectoderm

at the apical surface and the nucleus assumes a basal position. By the following cleavage (the 16-cell stage), there will be recognisable subpopulations of polar and apolar cells, and it is considered that these subpopulations furnish the inner cell mass and trophoblast, respectively. Biochemical and immunological differences exist between these two cell types. Table VI.1 is taken from Johnson (1979), and presents diverse features distinguishing the inner cell mass from the trophectoderm.

Such specialized features of early differentiation have almost certainly not yet been examined in any detail for the primate embryo. However, based largely on the outcome of in vitro fertilization and embryo culture, a concise review of histological features of the early human embryo is available (Mohr and Trounson 1984). Because in vitro fertilization can give rise to viable embryos, there is no reason to suppose that the in vitro organization of embryos that develop subsequently in vivo will differ significantly from that occurring spontaneously. If in vitro techniques involving microinjection of cells, nuclei or DNA sequences are one day to be applied to primate embryos, then experimental information concerning the first steps of differentiation would seem an important prerequisite.

REFERENCES

Adams CE (1958) Egg development in the rabbit: the influence of post coital ligation of the uterine tube and of ovariectomy. J Endocrinol 16:283–293
Adams CE (1973) The development of rabbit eggs in the ligated oviduct and their viability after re-transfer to recipient rabbits. J Embryol Exp Morph 29:133–144
Alden RH (1942) Aspects of the egg-ovary-oviduct relationship in the albino rat. J Exp Zool 90:159–170
Allen WR, Stewart F, Trounson AO, Tischner M, Bielanski W (1976) Viability of horse embryos after storage and long-distance transport in the rabbit. J Reprod Fertil 47:387–390

Asch RH, Fernandez EO, Magnasco LA, Pauerstein CJ (1978) Demonstration of a chorionic gonadotropin-like substance in rabbit morulae. Fertil Steril 29:444−446

Asch RH, Fernandez EO, Siler-Khodr TM, Pauerstein CJ (1979) Evidence for a human chorionic gonadotropin-like material in the rabbit blastocyst. Fertil Steril 32:697−703

Assheton R (1898) The development of the pig during the first ten days. Quart J Microscop Sci 41:329−359

Austin CR (1961) The mammalian egg. Blackwell Sci Publ, Oxford

Averill RLW, Adams CE, Rowson LEA (1955) Transfer of mammalian ova between species. Nature (Lond) 176:167

Beier HM, Lindner HR (1983) Fertilization of the human egg in vitro. Springer, Berlin Heidelberg New York

Betteridge KJ, Eaglesome MD, Flood PF (1979) Embryo transport through the mare's oviduct depends upon cleavage and is independent of the ipsilateral corpus luteum. J Reprod Fertil Suppl 27:387−394

Biggers JD (1981) In vitro fertilisation and embryo transfer in human beings. New Engl J Med 304:336−342

Biggers JD, Borland RM (1976) Physiological aspects of growth and development of the preimplantation mammalian embryo. Ann Rev Physiol 38:95−119

Biggers JD, Whittingham DG, Donahue RP (1967) The pattern of energy metabolism in the mouse oocyte and zygote. Proc Natl Acad Sci (Wash) 58:560−567

Borland RM, Hazra S, Biggers JD, Lechene CP (1977) The elemental composition of the environment of the gametes and preimplantation embryo during the initiation of pregnancy. Biol Reprod 16:147−157

Boué J, Boué A, Lazar P (1975) The epidemiology of human spontaneous abortions with chromosomal anomalies. In: Blandau RJ (ed) Ageing gametes, their biology and pathology. Karger, Basel, pp 330−348

Brinster RL (1965) Studies on the development of mouse embryos in vitro. II. The effect of energy source. J Exp Zool 158:59−68

Brinster RL (1973) Nutrition and metabolism of the ovum, zygote and blastocyst. In: Greep RO, Astwood EB (eds) Handbook of physiology, section 7, Endocrinology II. American Physiological Society, Washington, pp 165−185

Cavanagh AC, Morton H, Rolfe BE, Gidley-Baird AA (1982) Ovum factor: a first signal of pregnancy. Am J Reprod Immunol 2:97−101

Chang MC (1950) Development and fate of transferred rabbit ova or blastocyst in relation to the ovulation time of recipients. J Exp Zool 114:197−225

Chang MC (1969) Development of transferred ferret eggs in relation to the age of corpora lutea. J Exp Zool 171:459−464

Chang MC (1973) Effects of medroxyprogesterone acetate and of ethinyl oestradiol on the fertilization and transportation of ferret eggs. J Reprod Fertil 13:173−174

Chang MC, Pincus G (1951) Physiology of fertilisation in mammals. Physiol Rev 31:1−26

Corner GW (1942) The hormones in human reproduction. Princeton University Press

Craft I, McLeod F, Green S et al. (1982) Human pregnancy following oocyte and sperm transfer to the uterus. Lancet i:1031−1033

Donahue RP, Stern S (1968) Follicular cell support of oocyte maturation: production of pyruvate in vitro. J Reprod Fertil 17:395−398

Edmonds DK, Lindsay KS, Miller JF, Williamson E, Wood PJ (1982) Early embryonic mortality in women. Fertil Steril 38:447−453

Edwards RG (1964) Cleavage of one- or two-celled rabbit eggs in vitro after removal of the zona pellucida. J Reprod Fertil 7:413−415

Edwards RG (1980) Conception in the human female. Academic Press, London

Edwards RG, Purdy JM (1982) Human conception in vitro. Academic Press, London

Edwards RG, Steptoe PC, Purdy JM (1970) Fertilisation and cleavage in vitro of preovulatory human oocytes. Nature (Lond) 227:1307−1309

Fishel SB, Edwards RG, Purdy JM (1983) In vitro fertilization of human oocytes: factors associated with embryonic development in vitro, replacement of embryos, and pregnancy. In: Beier HM, Lindner HR (eds) Fertilization of the human egg in vitro. Springer, Berlin Heidelberg New York, pp 251−269

Greenwald GS (1958) Endocrine regulation of the secretion of mucin in the tubal epithelium of the rabbit. Anat Rec 130:477–495

Hansson A (1947) The physiology of reproduction in mink with special reference to delayed implantation. Acta Zool 28:1–136

Haour F, Saxena BB (1974) Detection of a gonadotropin in rabbit blastocyst before implantation. Science 185:444–445

Harper MJK, Pauerstein CJ (eds) (1976) Ovum transport and fertility regulation. Scriptor, Copenhagen

Hearn JP (1986) The embryo-maternal dialogue during early pregnancy in primates. J Reprod Fertil 76:809–819

Hearn JP, Gidley-Baird AA, O'Neill CT, Saunders PTK (1986) Platelet activating factor (PAF) as a monitor of preimplantation development in primates. Proc Soc Stud Fertil Abstract No 121:75

Hertig AT, Rock J, Adams EC (1956) A description of 34 human ova within the first 17 days of development. Am J Anat 98:435–494

Hunter GL, Bishop GP, Adams CE, Rowson LE (1962) Successful long-distance aerial transport of fertilized sheep ova. J Reprod Fertil 3:33–40

Hunter RHF (1974) Chronological and cytological details of fertilization and early embryonic development in the domestic pig *Sus scrofa*. Anat Rec 178:169–186

Hunter RHF (1977) Function and malfunction of the Fallopian tubes in relation to gametes, embryos and hormones. Europ J Obstet Gynecol Reprod Biol 7:267–283

Hunter RHF (1978) Intraperitoneal insemination, sperm transport and capacitation in the pig. Anim Reprod Sci 1:167–179

Hunter RHF, Fléchon B, Fléchon JE (1987) Pre- and peri-ovulatory distribution of viable spermatozoa in the pig oviduct: a scanning electron microscope study. Tissue & Cell 19:423–436

Johnson MH (1979) Intrinsic and extrinsic factors in preimplantation development. J Reprod Fertil 55:255–265

Johnson MH (1981) The molecular and cellular basis of preimplantation mouse development. Biol Rev 56:463–498

Kent HA Jr (1975) The two to four-cell embryos as source tissue of the tetrapeptide preventing ovulations in the hamster. Am J Anat 144:509–512

Lawson RAS, Rowson LEA, Adams CE (1972) The development of cow eggs in the rabbit oviduct and their viability after re-transfer to heifers. J Reprod Fertil 28:313–315

Leese HJ, Barton AM (1985) Production of pyruvate by isolated mouse cumulus cells. J Exp Zool 234:231–236

Leese HJ, Hooper MAK, Edwards RG, Ashwood-Smith MJ (1986) Uptake of pyruvate by early human embryos determined by a non-invasive technique. Human Reprod 1:181–182

McLaren A (1970) The fate of the zona pellucida in mice. J Embryol Exp Morph 23:1–19

McLaren A (1982) The embryo. In: Austin CR, Short RV (eds) Reproduction in mammals. 2, Embryonic and foetal development. Cambridge University Press, pp 1–25

Marston JH, Penn R, Sivelle PC (1977) Successful autotransfer of tubal eggs in the Rhesus monkey *Macaca mulatta*. J Reprod Fertil 49:175–176

Menezo Y, Laviolette P (1972) Les constituants aminés des sécrétions tubaires chez la lapine. Annls Biol Anim Biochim Biophys 12:383–396

Miller BG, Moore NW (1976) Effects of progesterone and oestradiol on RNA and protein metabolism in the genital tract and on survival of embryos in the ovariectomised ewe. Aust J Biol Sci 29:565–573

Mintz B (1962) Experimental study of the developing mammalian egg: removal of the zona pellucida. Science 138:594–595

Mohr L, Trounson A (1984) In vitro fertilisation and embryo growth. In: Wood C, Trounson A (eds) Clinical in vitro fertilisation. Springer, Berlin Heidelberg New York Tokyo Ch, 8, pp 99–115

Moor RM, Cragle RG (1971) The sheep egg: enzymatic removal of the zona pellucida and culture of eggs in vitro. J Reprod Fertil 27:401–409

Moor RM, Warnes GM (1978) Regulation of oocyte maturation in mammals. In: Crighton DB, Foxcroft GR, Haynes NB, Lamming GE (eds) Control of ovulation. Butterworths, London, pp 159–176

Moore NW, Adams CE, Rowson LEA (1968) Developmental potential of single blastomeres of the rabbit egg. J Reprod Fertil 17:527–531

Moore NW, Polge C, Rowson LEA (1969) The survival of single blastomeres of pig eggs transferred to recipient gilts. Aust J Biol Sci 22:979–982

Morton H, Hegh V, Clunie GJA (1976) Studies of the rosette inhibition test in pregnant mice: evidence of immunosuppression? Proc R Soc B 193:413–419

Morton H, Rolfe B, Clunie GJA, Anderson MJ, Morrison J (1977) An early pregnancy factor detected in human serum by the rosette inhibition test. Lancet i:394–397

Nancarrow CD, Wallace ALC, Grewal AS (1981) The early pregnancy factor of sheep and cattle. J Reprod Fertil Suppl 30:191–199

Oguri N, Tsutsumi Y (1972) Non-surgical recovery of equine eggs, and an attempt at non-surgical egg transfer in horses. J Reprod Fertil 31:187–195

O'Neill C (1985a) Thrombocytopenia is an initial maternal response to fertilisation in mice. J Reprod Fertil 73:559–566

O'Neill C (1985b) Examination of the causes of early pregnancy-associated thrombocytopenia in mice. J Reprod Fertil 73:567–577

O'Neill C (1985c) Partial characterisation of the embryo-derived platelet-activating factor in mice. J Reprod Fertil 75:375–380

O'Neill C, Saunders DM (1984) Assessment of embryo quality. Lancet ii:1035

O'Neill C, Pike IL, Porter RN, Gidley-Baird A, Sinosich MJ, Saunders DM (1985) Maternal recognition of pregnancy prior to implantation: methods for monitoring embryonic viability in vitro and in vivo. Annal NY Acad Sci 442:429–439

Ortiz ME, Bedregal P, Carvajal MI, Croxatto HB (1986) Fertilised and unfertilised ova are transported at different rates by the hamster oviduct. Biol Report 34:777–781

Pincus G (1936) The eggs of mammals. Exper Biol Monographs. Macmillan, New York

Polge C (1982) Embryo transplantation and preservation. In: Cole DJA, Foxcroft GR (eds) Control of pig reproduction. Butterworth Sci London, pp 277–291

Polge C, Adams CE, Baker RD (1972) Development and survival of pig embryos in the rabbit oviduct. Proc 7th Int Congr Anim Reprod Artif Insem, Munich 1:513–517

Pomeroy RW (1955) Ovulation and the passage of the ova through the Fallopian tubes in the pig. J Agric Sci 45:327–330

Psychoyos A (1960) Nouvelles contributions à l'étude de la nidation de l'oeuf chez la ratte. CR Hebd Séanc Acad Sci, Paris, D251:3073–3075

Psychoyos A (1973) Hormonal control of ovoimplantation. Vitams Horm 31:201–256

Rasweiler JJ (1979) Early embryonic development and implantation in bats. J Reprod Fertil 56:403–416

Roblero L, Biggers JD, Lechene CP (1976) Electron probe analysis of the elemental microenvironment of oviducal mouse embryos. J Reprod Fertil 46:431–434

Sauer MJ (1979) Review. Hormone involvement in the establishment of pregnancy. J Reprod Fertil 56:725–743

Sawicki JA, Magnuson T, Epstein CJ (1981) Evidence for expression of the paternal genome in the two-cell mouse embryo. Nature (Lond) 294:450–451

Short RV (1984) Species differences in reproductive mechanisms. In: Austin CR, Short RV (eds) Reproduction in mammals. 4, Reproductive fitness. Cambridge University Press, pp 24–61

Smart YC, Cripps AW, Clancy RL, Roberts TK, Lopata A, Shutt DA (1981) Detection of an immunosuppressive factor in human preimplantation embryo cultures. Med J Aust 1:78–79

Sreenan J, Scanlon P, Gordon I (1968) Culture of fertilised cattle eggs. J Agric Sci 70:183–185

Stone SL, Hamner CE (1975) Biochemistry and physiology of oviductal secretions. Gynecol Invest 6:234–252

Surani MAH, Barton SC, Norris ML (1987) Experimental reconstruction of mouse eggs and embryos: an analysis of mammalian development. Biol Reprod 36:1–16

Szöllösi D, Yotsuyanagi Y (1985) Activation of paternally derived regulatory mechanism in early mouse embryo. Dev Biol 111:256–259

Tervit HR, Rowson LEA (1974) Birth of lambs after culture of sheep ova in vitro for up to 6 days. J Reprod Fertil 38:177

Tervit HR, Whittingham DG, Rowson LEA (1972) Successful culture in vitro of sheep and cattle ova. J Reprod Fertil 30:493–497

Testart J, Frydman R (1985) Human in vitro fertilization. Elsevier, Amsterdam

Trounson AO, Moore NW (1974) The survival and development of sheep eggs following complete or partial removal of the zona pellucida. J Reprod Fertil 41:97−105

Van der Merwe M (1982) Histological study of implantation in the Natal clinging bat. J Reprod Fertil 65:319−323

Waites GT, Bell SC (1984) Identification of a murine pregnancy-associated protein (a-PAP) as a female-specific acute-phase reactant. J Reprod Fertil 70:581−589

Warnes GM, Moor RM, Johnson MH (1977) Changes in protein synthesis during maturation of sheep oocytes in vivo and in vitro. J Reprod Fertil 49:331−335

Whitten WK (1971) Nutrient requirements for the culture of preimplantation embryos in vitro. Adv Biosci 6:129−141

Whyte A, Heap RB (1983) Early pregnancy factor. Nature (Lond) 304:121−122

Willadsen SM (1979) A method for culture of micromanipulated sheep embryos and its use to produce monozygotic twins. Nature (Lond) 277:298−300

Willadsen SM (1981) The developmental capacity of blastomeres from 4- and 8-cell sheep embryos. J Embryol Exp Morph 65:165−172

Wilmut I, Sales DI, Ashworth CJ (1986) Maternal and embryonic factors associated with prenatal loss in mammals. J Reprod Fertil 76:851−864

Wintenberger-Torres S (1956) Les rapports entre l'oeuf en segmentation et la tractus maternel chez la brebis. Proc 3rd Int Congr Anim Reprod Cambridge 1:62−90

Wright RW, Bondioli KR (1981) Aspects of in vitro fertilisation and embryo culture in domestic animals. J Anim Sci 53:702−729

Wood C, Trounson A (eds) (1984) Clinical in vitro fertilization. Springer, Berlin Heidelberg New York Tokyo

CHAPTER VII

Transport of Embryos to the Uterus:
Normal and Abnormal Timing

CONTENTS

Introduction 127
Progression of Embryos Within the Fallopian Tube 128
Mechanisms Regulating Passage of Embryos 130
Pharmacological Disturbance of Egg Transport 133
Delayed Transport in the Aetiology of Tubal Pregnancy 135
References 136

Introduction

The fact that the timing of entry of embryos into the uterus from the Fallopian tubes is critical for subsequent survival in most species suggests that the mechanisms regulating this transport process must themselves be programmed with accuracy. Although the primate embryo may be a notable exception in this context of critical timing, evidence from farm and laboratory species indicates that premature or belated entry into the uterus compromises survival of the embryo. Some of the most persuasive findings are drawn from embryo transplantation studies, such as those of Chang (1950) and Adams (1979) in rabbits involving the use of donors and recipients at differing stages after ovulation. When seeking control systems that underlie timing of such remarkable sensitivity, it is worth recalling that (a) the egg is released from the ovary at approximately the start of corpus luteum formation − this gland being the principal ovarian source of progesterone − and (b) there is abundant evidence dating back to the 1930s indicating that the rate of passage of eggs or embryos from the tubes to the uterus can be grossly disturbed by systemic treatment with oestrogens or progestagens or, alternatively, by means of bilateral ovariectomy. In the first instance, therefore, a reasonable assumption might be that mechanisms that regulate passage of embryos to the uterus are strongly influenced by the prevailing secretion of ovarian steroid hormones.

From the above remarks, it will be apparent that inappropriate control mechanisms due, for example, to inadequate development of the corpus luteum or other sources of endocrine imbalance could constitute significant reasons leading to embryonic death. A clearer understanding of embryo transport mechanisms might therefore enable some causes of infertility to be alleviated

127

(Croxatto et al. 1979). Until recently, however, the reciprocal argument featured more widely in written reviews and especially in applications to grant-awarding authorities, i.e. that a disturbance imposed on the normal mechanisms of egg transport in the Fallopian tubes might be a valuable approach to contraception in women. In the light of evidence presented in this and subsequent chapters, such a viewpoint no longer seems fully tenable.

Progression of Embryos Within the Fallopian Tube

Specific remarks concerning passage of the unfertilized egg have been made in preceding chapters, especially in Chapter IV. As a generalization which seems to hold true for most mammals so far examined, the site of fertilization is at the ampullary-isthmic junction, and a rather rapid passage of newly-shed eggs occurs to this region of the tube (Fig. II.1). The ampullary-isthmic junction is less distinct in primates than, for example, in rabbits or pigs, and there is growing evidence that the primate egg lingers longer in the body of the ampulla (see Chap. IV; Pauerstein and Eddy 1979). But progress from the site of fertilization to the uterus is really the topic of this section, and it can be broadly summarized as follows.

The newly-penetrated egg remains in the vicinity of the site of fertilization for much of its tubal sojourn, and only progresses a substantial distance along the isthmus shortly before this change of environment occurs; in other words, a slow descent at first followed by a final rapid transit through the distal isthmus into the uterus. A reasonable interpretation here would be that the patency of the isthmus is gradually increasing under the overall influence of changing ovarian steroid secretion, and that progressive relaxation in the myosalpinx and mucosa enables ab-ovarian displacement of the embryos. When the extreme constriction of the isthmus that serves to control sperm transport in the pre-ovulatory phase is considered (Chaps. IV and V), and likewise the relative dimensions of eggs and spermatozoa, the change required in the patency of the isthmic lumen before eggs can pass is readily appreciated. A vital part of the strategy in regulating descent may be to avoid exposing embryos to reserves of viable spermatozoa still resident in the distal portion of the isthmus (Hunter 1977, 1987a).

As indicated, the above remarks are of a general nature, but there have been systematic studies in various mammals describing progress of embryos in the Fallopian tubes in relation to the interval elapsing from ovulation; for obvious reasons, the more abundant information concerns the polytocous species but none of it is comprehensive. This situation stems in part from technical problems, and especially reflects the difficulty of visualizing the denuded egg through the thick muscular coating of the isthmus. Accordingly, observations have invariably been post mortem or within in vitro culture systems, and thus there is always the possibility of artefacts. Nonetheless, a helpful viewpoint at this stage is that egg transport through the isthmus may take one of two forms, according to species: either a slow and more or less continuous progression or, alternatively, a relatively prolonged arrest at the ampullary-isthmic junction followed by a rapid progression to the uterus (Rousseau, Levasseur and Thibault 1987).

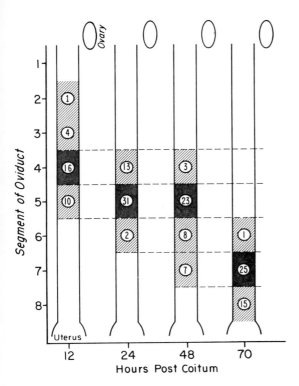

Fig. VII.1. Diagrammatic representation of the rabbit Fallopian tube to indicate the location of eggs at increasing intervals after ovulation. Note that most eggs remain close to the ampullary-isthmic junction for most of their tubal sojourn. (After Greenwald 1961)

In rabbits, the eggs are thought to be retained close to the ampullary-isthmic junction for much of their stay (Fig. VII.1), whilst the mucin layer is deposited around the zona pellucida, whereas passage through the distal isthmus into the uterus is rapid (Greenwald 1961). Entry of eggs into the uterus occurs approximately 72 h after ovulation. Turning to the large farm species, Assheton (1898) correctly inferred that pig eggs enter the uterus about 2 days after ovulation, although the erroneous figure of 3 days has since been quoted extensively (e.g. Corner 1921; Andersen 1927; Nalbandov 1964). However, the figure of 46–48 h for the tubal sojourn agrees well with the estimates of Pomeroy (1955) and Oxenreider and Day (1965); the latter authors provided evidence for a relatively slow initial descent of the isthmus and then a rapid transport through the terminal portion into the uterus. Although the time of entry of the sheep embryo into the uterus is approximately one day later, an essentially similar pattern of transport has been noted (Wintenberger 1953; Wintenberger-Torrès 1961), with the further suggestion that antiperistaltic activity of the isthmus for some 48–60 h after ovulation may contribute to the arrest of eggs at the ampullary-isthmic junction (Wintenberger-Torrès 1961). Hølst (1974) found that sheep eggs pass to the uterus some 66 h after ovulation, and that embryos recovered from the uterus at 68–72 h after ovulation were at the 8-cell stage. In cows, persuasive evidence exists for an arrest of eggs close to the ampullary-isthmic junction until shortly before descent to the uterus (Black and Davis 1962), at the stage of 8–16 cells. Whilst the period

129

of 5−7 days for passage of the horse embryo to the uterus is generally accepted, the actual form of progression and its consistency between animals require further definition.

Brief reference to the timing of egg entry into the uterus of certain primates is made in the reviews of Pauerstein and Eddy (1979) and Hearn (1986); they cite the customary range of 3−4 days after ovulation. Concerning egg transport in women, retention in the ampulla is apparently prolonged for up to 72 h, followed by a rapid transit through the isthmus to the uterus (Croxatto and Ortíz 1975; Croxatto et al. 1978). As suggested above, this pattern may function in part to avoid exposure of the embryo to relatively high concentrations of viable spermatozoa in the intra-mural portion of the tube (Hunter 1987b), and may be associated with a relatively tenacious column of mucus within the lumen of the isthmus (Jansen 1980). The human embryo is thought to have entered the uterus within 80 h of ovulation (Croxatto and Ortíz 1975; Pauerstein 1978; Pauerstein and Eddy 1979), probably at the 8−12-cell stage when mucus is no longer apparent in the lumen of the isthmus. A rather similar timing exists in the rhesus monkey and baboon (Eddy et al. 1975, 1976; Pauerstein and Eddy 1979). More extensive comparative data for diverse mammalian species are presented in the review of Wimsatt (1975).

Mechanisms Regulating Passage of Embryos

The principal mechanisms regulating the passage of embryos into the uterus involve activity of the myosalpinx and cilia, the reduction of oedema in the mucosa, and the direction of fluid flow in the tubal lumen. Some authors have also suggested a rôle for the mesenteries supporting the Fallopian tubes and uterus (Blandau 1969), but a directly-transmitted influence of contractions in the mesosalpinx and mesometrium on the transport of embryos is not easy to envisage. For the majority of species described in the literature, there is no clear distinction between the timing of entry of fertilized and unfertilized eggs into the uterus. Because this is not the situation in equids (Van Niekerk and Gerneke 1966; Betteridge and Mitchell 1974), various species of bat (Rasweiler 1979) and even golden hamsters (Ortíz et al. 1986), it raises the important question as to whether embryos can in some manner influence their own rate of passage to the uterus; the answer is increasingly becoming yes for an increasing number of species.

In the hours before ovulation in animals that exhibit a distinct behavioural oestrus (e.g. rabbits, pigs, sheep), waves of contraction in the isthmus proceed largely from the uterine end towards the ampullary-isthmic junction (i.e. in an adovarian direction). This predominant direction of muscular activity fades after ovulation and, by the time the embryo is scheduled to enter the uterus, spontaneous waves of contraction are seen to be essentially ab-ovarian. As an example, the study of Rodriguez-Martinez, Einarsson and Larsson (1982) involving pressure microtransducers placed in pig Fallopian tubes noted strong peristaltic waves of contraction in the isthmus at the time when eggs would be passing to the uterus. This would therefore suggest an active rôle of smooth muscle in displacing the eggs, at least in this portion of the duct system.

130

The muscular coats constrict the tubal lumen under the dominance of pre-ovulatory oestrogen secretion from the Graafian follicle but, as the concentration of circulating progesterone increases, so the myosalpinx becomes progressively more relaxed. One means of physiological control here involves the very rich adrenergic innervation of the isthmus (Brundin 1965). α-Adrenergic receptors in the smooth muscle cells are potentiated under the influence of oestrogens and promote contraction of the myosalpinx, especially in the layers of the circular coat, thereby providing a form of physiological sphincter. β-Adrenergic receptors, by contrast, are potentiated under increasing titres of progesterone, and they in turn promote gradual relaxation of the smooth muscle coat. Thus, critical events in the wall of the tube are being regulated by the change from pre-ovulatory oestrogen dominance to one of post-ovulatory progesterone (see Hunter 1977). Moreover, in the context of regulating the passage of eggs or embryos through the isthmus, the very rich adrenergic innervation of this portion of the tube enables it to act incisively in the manner just described.

Analysis of precise interactions between the prevailing balance of circulating steroid hormones and the content of available catecholamines in the wall of the tube is still awaited. However, limited information does exist on steroid hormone interactions with the prostaglandin content of Fallopian tube tissues which may influence the pattern of contractile activity, probably by release at nerve terminals (Harper et al. 1980). In simplified terms, prostaglandins of the F series are thought to promote contraction in the isthmus under oestrogen dominance, whereas those of the E series lead to relaxation and are activated under pro-gesterone dominance. Oestradiol treatment significantly increases the PGF levels in the wall of the tube whereas progesterone treatment depresses them to basal levels in rabbits (Saksena and Harper 1975).

This proposed rôle of tissue prostaglandins is open to debate, since there is some contradiction between the action of specific prostaglandins on tubal func-tion in vivo and that recorded in vitro, and likewise due to differential effects of specific prostaglandins on the longitudinal and circular layers of muscle. Nonetheless, prostaglandins of the F series may assist retention of eggs in the isthmus by virtue of their occlusive action, whereas those of the E series would facilitate passage by diminishing the occlusive effects (Spilman and Harper 1973; Spilman 1976). The relationships between these two prostaglandins in regulating egg transport in rabbits have been discussed (Rajkumar, Garg and Sharma 1979). There may be regional effects of prostaglandins within the isthmus, with the uterine quarter tending to contract whilst the remaining three-quarters relax. In both rabbits and humans, however, prostaglandin $F_{2\alpha}$ activates tubal contractility while prostaglandin E induces relaxation (Maia, Barbosa and Coutinho 1976; Barbosa et al. 1980). Overall, therefore, prostaglandins of the F series work in conjunction with α-adrenergic activation to constrict the smooth muscle coat of the tube and reduce the dimensions of the lumen during the pre-ovulatory phase of oestrogen secretion, a situation that is reversed as the corpus luteum develops and its secretion of progesterone increases.

Tracts of cilia are conspicuous in the isthmus of only a few mammals (Fig. VII.2), but they have been clearly documented for rabbits and pigs (Blandau and Gaddum-Rosse 1974; Wu, Carlson and First 1976; Fléchon and Hunter 1981),

Fig. VII.2. A scanning electron micrograph of a portion of the isthmus prepared from a pig Fallopian tube to illustrate the extensive tracts of cilia on the surface of the longitudinal folds. (Adapted from Hunter, Fléchon and Fléchon 1987)

and to a lesser extent in women (Ferenczy et al. 1972; Patek 1974; Jansen 1980). As noted in Chapter IV, the cilia that line the isthmus of pigs and rabbits beat principally towards the site of fertilization in the pre-ovulatory interval and are thought to assist in the passage of spermatozoa. Thereafter, they may also assist the transport of embryos to the uterus, both directly by means of contact with the egg surface and indirectly by propagating specific currents of fluid flow (Chap. III). Transport of embryos through the isthmus of cows and sheep may also be aided by cilia which, in these species, are said invariably to beat towards the uterus (Gaddum-Rosse and Blandau 1976), and a similar situation has been reported for the human isthmus (Edwards 1980). But the extent to which tubal cilia play a vital rôle in women has been questioned from several lines of evidence, not least the finding that patients with the 'immotile cilia syndrome' may be fertile (Pauerstein and Eddy 1979). But this observation has received a qualifying comment in Chapter IV. Even so, the effectiveness of cilia in promoting egg transport will presumably be related to the size of the tubal lumen, which will reflect not only the condition of the myosalpinx but also the extent of oedema in the prominent and much-folded mucosa (Fig. VII.3). The condition of this layer of the tube is again an expression of the prevailing balance of gonadal steroids and, in the domestic pig, for example, it is extremely oedematous at oestrus (Andersen

132

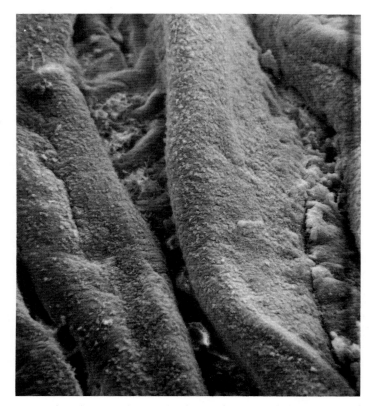

Fig. VII.3. A scanning electron micrograph of a portion of the isthmus prepared from a pig Fallopian tube close to the time of ovulation showing the prominent condition of the longitudinal folds due to oedema in the much-folded mucosa. (Adapted from Fléchon and Hunter 1981)

1928). This oedema is markedly reduced by the time of egg passage into the uterus, and can be reduced by systemic (Day and Polge 1968) or, more specifically, by local microinjections (Hunter 1972) of a solution of progesterone in oil.

Overall, therefore, the regulatory influence of ovarian steroids on the process of egg transport appears to be exerted mainly through the excitability and conduction properties of oviductal smooth musculature and the morphology of the mucosa.

Pharmacological Disturbance of Egg Transport

An observation of long standing in laboratory animals is that the expected rate of progress of eggs through the Fallopian tubes to the uterus can be disrupted in the presence of anomalous levels of ovarian steroid hormones, be these endogenous or exogenous in origin (review by Bennett 1970). This disruption may be in the form of 'tube locking' of eggs, as upon oestrogen administration in mice, hamsters, guinea pigs and rabbits (Burdick and Pincus 1935; Greenwald 1967; Humphrey and Martin 1968) or, alternatively, an accelerated tubal passage of eggs to the uterus following progestagen administration in rabbits (e.g. Chang 1966). Both forms of anomaly compromise development of the embryo and tend to be lethal in farm animals and laboratory rodents. As noted above, the time of entry into the uterus is critical for subsequent development and implantation in most mammals examined. Historical evidence for an influence of steroid hormones on the contractile activity of the Fallopian tubes dates at least from the studies of Corner (1923) and Seckinger (1923). More direct demonstrations of steroid hormones perturbing tubal egg transport come from the induction of ovulation in the presence of active corpora lutea (Wislocki and Snyder 1933; Hunter 1967) or specific injection of steroid hormones close to the time of ovulation (Day and Polge 1968). Passive immunization against ovarian steroids can also be used to clarify the contribution of individual steroid hormones to the process of egg transport (Bigsby, Duby and Black 1986).

Across species, it is difficult to state categorically whether exogenous oestrogens will cause 'tube locking' and progesterone an acceleration of descent. This is not least since the underlying processes are influenced by (1) the dose of hormone (which is invariably pharmacological), (2) the actual oestrogenic or progestational compound administered, and (3) most critically, the time relationship between treatment and ovulation. As a general rule, however, oestrogen treatment retards tubal transport of eggs, whereas progesterone treatment facilitates a more rapid descent. But treatment with progesterone may only cause such a response when acting on oestrogen-primed tissues of the tube (Boling and Blandau 1971). A further point of detail is that whilst 'tube locking' of eggs is a classical response to appropriate doses of oestrogens in mice, hamsters and guinea pigs, it could not be demonstrated in rats (Greenwald 1967).

As would be expected from remarks in the preceding section of this chapter, systemic administration of various prostaglandins can also be used to disturb egg or embryo transport to the uterus, although it must be questioned whether such a result is obtained solely by a direct action on the Fallopian tube or if the pattern of ovarian steroid secretion is modified as a consequence. Nonetheless, prostaglandin $F_{2\alpha}$ has been reported to accelerate egg transport in rabbits (Chang and Hunt 1972; Ellinger and Kirton 1974; Spilman et al. 1976), whereas inhibitors of prostaglandin synthesis such as indomethacin may act to retard the initial stages of egg transport (Hodgson 1976). Endotoxin likewise accelerates egg transport in rabbits if administered close to the time of ovulation, and this influence has been shown to be prostaglandin-mediated and to be reversible by indomethacin (Harper et al. 1981). Turning to primates, there may be species differences in the response to prostaglandins, although a comprehensive range of studies has yet to be completed. Even so, prostaglandin $F_{2\alpha}$ acts to increase tubal contractility in marmoset monkeys and women, especially shortly after ovulation, although its action in rhesus monkeys may be more sensitive to the stage of the menstrual cycle (Spilman 1974; Barbosa et al. 1980). Whether treatment with acceptable doses of prostaglandins will lead consistently to accelerated egg transport in women remains under debate, for Croxatto et al. (1979) and Pauerstein and Eddy (1979) have noted that stimulation of tubal contractility does not necessarily accelerate egg transport. There is the additional consideration that such a procedure may not be contraceptive in its effect, an aspect which is discussed in Chapters VIII and IX.

Again, drugs that modify the function of the autonomic nervous system, such as a- or β-adrenergic blockers, might be expected to alter normal tubal function so that egg transport is blocked or accelerated. However, pharmacological interference with the adrenergic system has been stated not to disturb egg transport in several species of mammal (Pauerstein and Eddy 1979), indicating − along with evidence from microsurgical studies (Chap. VIII) − that such innervation is not a primary control mechanism for regulating egg passage through the tubes.

Preliminary data bearing on interference with neuro-endocrine programming of human Fallopian tubes were briefly reviewed by Croxatto et al. (1979). Stimulation of β-adrenergic receptors was known to relax the circular muscle of the tubal isthmus but ritodrine, a β-adrenergic agonist given orally at a dose of 10 mg, had little detectable influence on the myosalpinx in terms of egg transport (Coutinho, Maia and Filho 1974). Pharmacological approaches of this kind in women would seem to be fraught with difficulty and/or danger, not least because of the risk of undesirable side effects for the patient − or indeed for the embryo should the treatment prove unsuccessful as a contraceptive measure. It is at this point that the value of non-human primate models is traditionally urged in experimental programmes but, ultimately, a critical decision has to be made as to whether to commence specific trials in women.

Delayed Transport in the Aetiology of Tubal Pregnancy

Whilst the subject of ectopic pregnancy is discussed more extensively in Chapter VIII, it is worth considering briefly here whether treatments that modify the rate of egg transport in women might promote an enhanced incidence of tubal pregnancies. Unfortunately, there is no satisfactory animal model in which to assess the risk of an ectopic pregnancy. The evidence linking modified rates of egg transport with tubal pregnancy is equivocal but should be taken into account, not least before embarking on further clinical trials that aim to modify the rate of tubal egg transport as a possible approach to contraception. It is of some relevance that in a series of ten patients treated with oestrogens as a possible post-coital contraceptive, Coutinho (1971) reported two ectopic pregnancies. Morris and van Wagenen (1973) reported a 10% incidence of ectopic pregnancy after post-ovulatory treatment with oestrogens. Endocrine imbalances, possibly associated with an inadequate development of the corpus luteum and secretion of progesterone, may be sufficient to disrupt the normal passage of the embryo into the uterus. Of course, the contribution of 'spontaneous' tubal spasm to this condition remains unknown, but presumably such spasms could be associated with endocrine and/or psychosomatic disturbances.

A related hypothesis has been offered by Rousseau et al. (1987). They argue that in species such as our own showing discontinuous transport of the egg through the Fallopian tube, peristaltic waves of contraction are required to displace the embryo along the isthmus to the uterus. A diminished intensity and/or frequency of peristalsis might therefore underlie the elevated proportion of ectopic pregnancies found in the ampulla compared with the isthmus (Chap. VIII), or indeed induced after non-surgical introduction of human embryos into the uterus on Day 2 (Chap. IX).

Whilst such speculation is of considerable interest, it also serves to emphasize that much remains to be learnt of the interacting factors that promote transport of the human embryo along the Fallopian tube. If the embryo usually acts locally to influence its own displacement, then a delayed or defective release of components from the zygote could well be involved in the aetiology of tubal pregnancy under some circumstances. Such reasoning would be strengthened if embryos located ectopically were shown to be intrinsically abnormal, perhaps — for example — containing a preponderance of tissues derived from the syncytiotrophoblast.

REFERENCES

Adams CE (1979) Consequences of accelerated ovum transport, including a re-evaluation of Estes' operation. J Reprod Fertil 55:239–246
Andersen DH (1927) The rate of passage of the mammalian ovum through various portions of the Fallopian tube. Am J Physiol 82:557–569
Andersen DH (1928) Comparative anatomy of the tubo-uterine junction. Histology and physiology in the sow. Am J Anat 42:255–305
Assheton R (1898) The development of the pig during the first ten days. Quart J Microscop Sci 41:329–359
Barbosa IC, Maia H, Dourado V, Coutinho E (1980) Effect of prostaglandin $F_{2\alpha}$ on oviduct contractility in marmoset monkeys. Fertil Steril 33:197–200
Bennett JP (1970) The effect of drugs on egg transport. Adv Biosci 4:165–178
Betteridge KJ, Mitchell D (1974) Direct evidence of retention of unfertilized ova in the oviduct of the mare. J Reprod Fertil 39:145–148
Bigsby RM, Duby RT, Black DL (1986) Effects of passive immunization against estradiol on rabbit ovum transport. Int J Fertil 31:240–245
Black DL, Davis J (1962) A blocking mechanism in the cow oviduct. J Reprod Fertil 4:21–26
Blandau RJ (1969) Gamete transport — comparative aspects. In: Hafez ESE, Blandau RJ (eds) The mammalian oviduct. University of Chicago Press, pp 129–162
Blandau RJ, Gaddum-Rosse (1974) Mechanism of sperm transport in pig oviducts. Fertil Steril 25:61–67
Boling JL, Blandau RJ (1971) Egg transport through the ampullae of the oviducts of rabbits under various experimental conditions. Biol Reprod 4:174–184
Brundin J (1965) Distribution and function of adrenergic nerves in the rabbit Fallopian tube. Acta Physiol Scand 66, Suppl 259:1–57
Burdick HO, Pincus G (1935) The effect of oestrin injections upon the developing ova of mice and rabbits. Am J Physiol 111:201–208
Chang MC (1950) Development and fate of transferred rabbit ova or blastocyst in relation to the ovulation time of recipients. J Exp Zool 114:197–225
Chang MC (1966) Effects of oral administration of medroxyprogesterone acetate and ethinyl estradiol on the transportation and development of rabbit eggs. Endocrinology 79:939–948
Chang MC, Hunt DM (1972) Effect of prostaglandin $F_{2\alpha}$ on the early pregnancy of rabbits. Nature (Lond) 236:120–121
Corner GW (1921) Cyclic changes in the ovaries and uterus of the sow and their relation to the mechanism of implantation. Contrib Embryol 13:117–145
Corner GW (1923) Cyclic variation in uterine and tubal contraction waves. Am J Anat 32:345–351

Coutinho EM (1971) Tubal and uterine motility. In: Diczfalusy E, Borell U (eds) Control of human fertility. Nobel Symposium 15, Wiley, New York, pp 97–115

Coutinho EM, Maia H, Filho JA (1974) The inhibitory effect of ritodrine on human tubal activity in vivo. Fertil Steril 25:596–601

Croxatto HB, Ortiz MES (1975) Egg transport in the Fallopian tube. Gynecol Invest 6:215–225

Croxatto HB, Ortiz MES, Diaz S, Hess R, Balmaceda J, Croxatto HD (1978) Studies on the duration of egg transport by the human oviduct. II. Ovum location at various intervals following luteinizing hormone peak. Am J Obstet Gynecol 132:629–634

Croxatto HB, Ortiz MES, Diaz S, Hess R (1979) Attempts to modify ovum transport in women. J Reprod Fertil 55:231–237

Day BN, Polge C (1968) Effects of progesterone on fertilisation and egg transport in the pig. J Reprod Fertil 17:227–230

Eddy CA, Garcia RG, Kraemer DC, Pauerstein CJ (1975) Detailed time course of ovum transport in the Rhesus monkey (*Macaca mulatta*). Biol Reprod 13:363–369

Eddy CA, Turner T, Kraemer D, Pauerstein CJ (1976) Pattern and duration of ovum transport in the baboon (*Papio anubis*). Obstet Gynecol 47:658–664

Edwards RG (1980) Conception in the human female. Academic Press, London

Ellinger JV, Kirton KT (1974) Ovum transport in rabbits injected with prostaglandin E_1 and $F_{2\alpha}$. Biol Reprod 11:93–96

Ferenczy A, Richart RM, Agate FJ Jr, Purkerson ML, Dempsey EW (1972) Scanning electron microscopy of the human Fallopian tube. Science 175:783–784

Fléchon JE, Hunter RHF (1981) Distribution of spermatozoa in the utero-tubal junction and isthmus of pigs and their relationship with the luminal epithelium after mating: a scanning electron microscope study. Tissue & Cell 13:127–139

Gaddum-Rosse P, Blandau RJ (1976) Comparative observations on ciliary currents in mammalian oviducts. Biol Reprod 14:605–609

Greenwald GS (1961) A study of the transport of ova through the rabbit oviduct. Fertil Steril 12:80–95

Greenwald GS (1967) Species differences in egg transport in response to exogenous oestrogen. Anat Rec 157:163–172

Harper MJK, Coons LW, Radicke DA et al. (1980) Role of prostaglandins in contractile activity of the ampulla of the rabbit oviduct. Am J Physiol 238:157–166

Harper MJK, Norris CJ, Friedrichs WE, Moreno A (1981) Poly I:C accelerates ovum transport in the rabbit by a prostaglandin-mediated mechanism. J Reprod Fertil 63:81–89

Hearn JP (1986) The embryo-maternal dialogue during early pregnancy in primates. J Reprod Fertil 76:809–819

Hodgson BJ (1976) Effects of indomethacin and ICI 46474 administered during ovum transport on fertility in rabbits. Biol Reprod 14:451–457

Hølst PJ (1974) The time of entry of ova into the uterus of the ewe. J Reprod Fertil 36:427–428

Humphrey K, Martin L (1968) The effect of oestrogen and anti-oestrogens on ovum transport in mice. J Reprod Fertil 15:191–197

Hunter RHF (1967) Polyspermic fertilization in pigs during the luteal phase of the oestrous cycle. J Exp Zool 165:451–460

Hunter RHF (1972) Local action of progesterone leading to polyspermic fertilization in pigs. J Reprod Fertil 31:433–444

Hunter RHF (1977) Function and malfunction of the Fallopian tubes in relation to gametes, embryos and hormones. Europ J Obstet Gynecol Reprod Biol 7:267–283

Hunter RHF (1987a) Peri-ovulatory physiology of the oviduct, with special reference to progression, storage and capacitation of spermatozoa. In: Mohri H (ed) New horizons in sperm cell research. Jpn Sci Soc, Tokyo, pp 31–45

Hunter RHF (1987b) Human fertilisation in vivo, with special reference to progression, storage and release of competent spermatozoa. Human Reprod 2:329–332

Hunter RHF, Fléchon B, Fléchon JE (1987) Pre- and peri-ovulatory distribution of viable spermatozoa in the pig oviduct: a scanning electron microscope study. Tissue & Cell 19:423–436

Jansen RPS (1980) Cyclic changes in the human Fallopian tube isthmus and their functional importance. Am J Obstet Gynecol 136:292–308

Maia H, Barbosa I, Coutinho EM (1976) Relationship between cyclic AMP levels and oviductal contractility. In: Harper MJK, Pauerstein CJ (eds) Symposium on ovum transport and fertility regulation. Scriptor, Copenhagen, pp 168–181

Morris JM, van Wagenen G (1973) Interception: the use of postovulatory estrogens to prevent implantation. Am J Obstet Gynecol 115:101–106

Nalbandov AV (1964) Reproductive physiology, 2nd edn. Freeman, San Francisco

Ortíz ME, Bedregal P, Carvajal MI, Croxatto HB (1986) Fertilised and unfertilised ova are transported at different rates by the hamster oviduct. Biol Reprod 34:777–781

Oxenreider SL, Day BN (1965) Transport and cleavage of ova in swine. J Anim Sci 24:413–417

Patek E (1974) The epithelium of the human Fallopian tube. Acta Obstet Gynecol Scand 53, Suppl 31:1–28

Pauerstein CJ (1978) From Fallopius to fantasy. Fertil Steril 30:133–140

Pauerstein CJ, Eddy CA (1979) The role of the oviduct in reproduction; our knowledge and our ignorance. J Reprod Fertil 55:223–229

Pomeroy RW (1955) Ovulation and the passage of the ova through the Fallopian tubes in the pig. J Agric Sci 45:327–330

Rajkumar K, Garg SK, Sharma PL (1979) Relationship between concentration of prostaglandins E and F in the regulation of ovum transport in rabbits. Prostaglandins Med 2:445–454

Rasweiler JJ (1979) Differential transport of embryos and degenerating ova by the oviducts of the long-tongued bat, *Glossophaga soricina*. J Reprod Fertil 55:329–334

Rodriguez-Martinez H, Einarsson S, Larsson B (1982) Spontaneous motility of the oviduct in the anaesthetized pig. J Reprod Fertil 66:615–624

Rousseau JP, Levasseur MC, Thibault C (1987) L'isthme et la jonction utéro-tubaire. Contracept Fertil Sex 15:227–239

Saksena SK, Harper MJK (1975) Relationship between concentration of prostaglandin F (PGF) in the oviduct and egg transport in rabbits. Biol Reprod 13:68–76

Seckinger DL (1923) Spontaneous contractions of the Fallopian tube of the domestic pig with reference to the oestrous cycle. Bull John Hopk Hosp 34:236–239

Spilman CH (1974) Oviduct motility in the rhesus monkey: spontaneous activity and response to prostaglandins. Fertil Steril 25:935–939

Spilman CH (1976) Prostaglandins, oviductal motility and egg transport. In: Harper MJK, Pauerstein CJ (eds) Symposium on ovum transport and fertility regulation. Scriptor, Copenhagen, pp 197–211

Spilman CH, Harper MJK (1973) Effect of prostaglandins on oviduct motility in estrous rabbits. Biol Reprod 9:36–45

Spilman CH, Beuving DC, Roseman TJ, Larion LJ (1976) Effect of vaginally administered 15(S)-15-methyl-$PGF_{2\alpha}$ on egg transport and fertility in rabbits. Proc Soc Exp Biol Med 151:575–578

Van Niekerk CH, Gerneke WH (1966) Persistence and parthenogenetic cleavage of tubal ova in the mare. Onderstepoort J Vet Res 33:195–231

Wimsatt WA (1975) Some comparative aspects of implantation. Biol Reprod 12:1–40

Wintenberger S (1953) Recherches sur les relations entre l'oeuf et le tractus maternel pendant les premiers stades du développement chez les mammifères. Etude de la traversée de l'oviducte par l'oeuf féconde de brebis. Annls Zootech 2:269–273

Wintenberger-Torrès S (1961) Mouvements des trompes et progression des oeufs chez la brebis. Annls Biol Anim Biochim Biophys 1:121–133

Wislocki GB, Snyder FF (1933) The experimental acceleration of the rate of transport of ova through the Fallopian tube. Bull Johns Hopk Hosp 52:379–386

Wu ASH, Carlson SD, First NL (1976) Scanning electron microscopic study of the porcine oviduct and uterus. J Anim Sci 42:804–809

CHAPTER VIII

Malfunction of the Fallopian Tubes: Spontaneous Conditions and Surgical Studies

CONTENTS

Introduction 139
Anomalies of the Fallopian Tubes 140
Ectopic (Tubal) Pregnancy 141
Intra-Uterine Fertilization and Estes' Operation 145
Tubal Resection and Reconstructive Surgery 149
Microsurgery and Subsequent Pregnancy 153
References 156

Introduction

Much of the text so far has considered normal physiological processes in the Fallopian tubes, especially as they concern progression and fusion of the gametes and the first steps in development of the embryo. Whilst the emphasis is thereby – and perhaps appropriately – placed on the condition of fertility, problems in the female duct system that are associated with infertility, if not complete sterility, cannot be overlooked. The intention of this chapter is not to divert the reader from perusing suitable clinical texts on many of these matters, but rather to offer informed comment on clinical topics in the light of studies in experimental animals. It goes without saying that the scope for relevant experimentation is greater in domestic animals than in our own species, and the same is true in terms of the latitude for examining embryos to establish morphological normality and potential viability. Some of the paragraphs that follow assume a historical perspective, not only in their own right but also in view of the material contained in Chapter IX. However, it would be rash to predict the response of future societies to technical advances in vitro, and therefore the limitations of earlier techniques are certainly worth reviewing.

Involuntary tubal infertility remains a major problem in women. The following figures may help to give a sense of proportion; they are taken from the review of Fergusson (1982). Tubal factors account for approximately 20% of all cases of infertility and, of these, 70% remain incurable. In round figures, this means, for example, that 20000 women in the United Kingdom and 90000 women in the United States, will suffer from permanent tubal sterility.

Anomalies of the tubes have frequently been documented in domestic animals and humans, and detailed descriptions are to be found in appropriate clinical texts (e.g. Novak and Rubin 1952; Nalbandov 1964; Novak and Woodruff 1967; Woodruff and Pauerstein 1969; Laing 1979; Chamberlain and Winston 1982). The range of anomalies is wide, and extends from a congenital absence of the tubes or portions thereof to major diverticula in the ducts and, on occasions, to duplication of individual tubes; bizarre deformities have been noted in all portions of the ducts, as well as occlusion at all levels. In addition to congenital problems, there are conditions that arise during the reproductive lifespan such as hydrosalpinx, fimbrial, peri-tubal and peri-ovarian adhesions, and even gross kinking or torsion of the tube(s), again sometimes associated with adhesions. These and other adverse conditions may also develop as a sequel to pelvic surgery. It is not helpful to generalize here on the extent to which treatment or remedial surgery will be beneficial in individual instances, although it is worth remarking that most of the above conditions in women have been subjected to some form of intervention. A new light is thrown on such problems by advances in the techniques of laparoscopy and egg recovery, and the associated procedures of in vitro fertilization and embryo transplantation.

Pathology is not the subject of this text but a brief comment will not be out of place. Inflammatory diseases are the most important cause of tubal disorders because of their frequency and the probability of serious consequences. Moreover, the anatomical siting of the Fallopian tubes between the ovaries and uterus renders them susceptible to the diseases and afflictions of both (Novak and Rubin 1952). Inflammatory disorders of the uterus may frequently extend into the tubes, for any ad-uterine current caused by the tubal cilia is a weak means of protection, and the narrow lumen of the intra-mural portion is itself an ineffective barrier. A related problem is that swelling of the mucosa in the region of the utero-tubal junction can completely occlude the duct lumen, thereby preventing drainage of inflammatory secretions into the uterine cavity. The diverse forms and consequences of salpingitis are well illustrated in the major review of Novak and Rubin (1952).

Perhaps less widely known, but of considerable research interest, is the condition of underdeveloped Fallopian tubes associated with differing degrees of intersexuality. The presence of testis-like tissue in the gonad(s) of genetic females may lead to poorly-developed or even vestigial Fallopian tubes (which are proximal portions of the Müllerian duct system − see Chap. II); proliferation of the neighbouring male duct as an epididymis (part of the Wolffian system) is usually found in this situation. Recent studies on intersexuality in pigs have highlighted such anomalies and focussed on their possible aetiology (Hunter, Baker and Cook 1982; Hunter, Cook and Baker 1985). Whereas testicular tissue in an XX ovotestis does not show spermatogenesis or even contain detectable germ cells, Sertoli-like cells are abundant and apparently synthetically active. It is these somatic cells of the seminiferous tubules that secrete Müllerian-inhibiting hormone (Josso et al. 1979; Vigier et al. 1984), which acts locally to inhibit development of the Fallopian tube whilst Leydig cell androgens promote elaboration of the epididymal duct. Although the uterus shows normal development, the isthmus of the tube is poorly defined and the ampulla scarcely visible macroscopically (Fig. II.4). Such intersex animals are now proving invaluable for experimental studies on modifications to the duct systems.

Ectopic (Tubal) Pregnancy

As is widely known in clinical circles, potential sites for an ectopic pregnancy are not restricted to the Fallopian tubes proper; human embryonic tissue may also implant and embark upon a vigorous form of development in the distal extremity of the intramural portion of the tube, or in ovarian tissue, or in other abdominal sites (Benirschke 1969). However, this section of the chapter will be specifically concerned with development of embryos within the tube (Fig. VIII.1), and especially with the question as to why ectopic pregnancies are apparently peculiar to primates. Attempts to develop non-primate models of extra-uterine pregnancy have not met with success, whilst the reported incidence of this condition in sub-human primates is extremely low (Adams 1977).

The scope for intra-tubal development of the primate embryo seems to be limited principally by the precise site of implantation. There is clearly more scope for development in the proximal portion of the ampulla than in the distal portion of the isthmus. Over and above the relative dimensions of these portions of the tube, there is a different potential for stretching the musculature. Flexibility exists in the ampulla to a significantly greater degree than in the isthmus, due to the much thicker muscular wall of the latter. This difference in the ability to stretch can be demonstrated in post-operative samples simply by injecting liquid through the fimbriated end and occluding both extremities: the ampullary portion will expand very considerably before rupturing. Nonetheless, short of clinical intervention or spontaneous death of the embryo, the most probable sequel to ectopic implantation is rupture of the wall of the tube irrespective of precisely where devel-

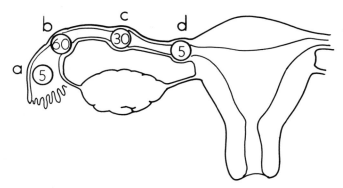

Fig. VIII.1. Diagrammatic representation of the genital tract to indicate the more commonly observed location and incidence of tubal ectopic pregnancies. The numbers refer to the percentage of such pregnancies in the fimbriated extremity of the tube (*a*), ampulla (*b*), isthmus (*c*) and the intra-mural portion (*d*). (Courtesy of Prof. C. J. Pauerstein)

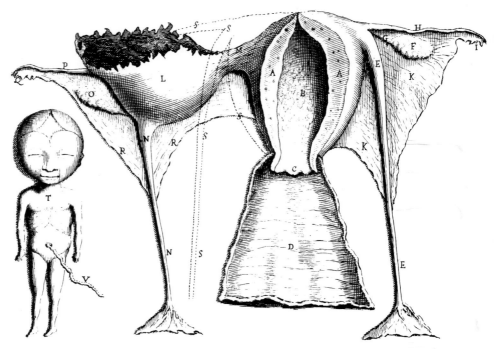

Fig. VIII.2. A classical depiction of an ectopic tubal pregnancy appearing in Mauriceau's (1675) Traité des Maladies des Femmes Grosses. (Courtesy of the Bibliothèque Nationale, Paris)

opment commenced; the structure of the wall is much weakened by the very invasiveness of the trophoblast. The associated problems of pain, haemorrhage, vascular collapse, shock and fainting are grave. Indeed, the 1979–1981 Report on Confidential Enquiries into Maternal Deaths in England and Wales noted that ectopic pregnancy accounted for 11.4% of all maternal deaths, being the fourth major cause after hypertensive diseases, pulmonary embolism and anaesthesia (Randall 1986).

As to the stage of gestation that can be attained by a human tubal pregnancy, there are well-documented records of 3 months or beyond dating from the nineteenth and early twentieth century but haemorrhagic ablation of the placenta or rupture of the tube, or both, usually occur in the first 2 to 3 months (Benirschke 1969). More advanced pregnancies (Fig. VIII.2) were found before the advent of modern intensive medicine and sensitive radioimmuno-assays for pregnancy hormones, especially those for human chorionic gonadotrophin. However, on most occasions, development is primarily of extra-embryonic tissues rather than of the embryo itself. All clinicians would earnestly hope for an early diagnosis, not only to protect the life of the patient but also to avoid irreversible damage to the tube itself. Apart from the necessary skills and alertness, diagnosis can be assisted by the techniques of ultrasonic scanning and/or laparoscopy. A majority of deaths from ectopic pregnancy (64%) in the Report referred to above were considered avoidable, the most frequent problem being a delayed detection.

142

Conditions that promote the occurrence of ectopic pregnancy would seem most logically to arise from those that retard passage of embryos to the uterus. Pelvic infection is traditionally viewed as a starting point, with salpingitis due to *Chlamydia trachomatis* being a principal cause of progressive tubal occlusion (Henry-Suchet and Loffredo 1980; Sweet 1982; Brunham et al. 1985; Tuffrey et al. 1986). Gonococcal and tuberculous infection may also be involved, and complications can arise following septic abortions. Scarring and subsequent occlusion of the tube tend to be frequent sequelae to the inflammation arising from bacterial infection. A recent review suggests that about 15% of women will become sterile after one episode of salpingitis (30% after two infections), 15% will develop chronic pain necessitating subsequent surgery, and the incidence of ectopic pregnancy in those who do conceive will be ten times greater than in women without infection (Eschenbach 1985). In this context, Pauerstein (1974) also mentions the presence of diverticula in both the ampulla and isthmus of the tube, i.e. salpingitis isthmica nodosa. The structures were thought to arise as a sequel to disease, and may have acted to trap the descending embryo. But Pauerstein states that no gross or microscopic evidence of salpingitis was found in a majority of women who suffer ectopic pregnancy. Nevertheless, he concedes that earlier episodes of infection may have acted adversely to modify the tubal epithelium. The condition of the ciliated epithelium should certainly be questioned.

Somewhat paradoxically, two completely different means of contraception may be associated with an enhanced incidence of ectopic pregnancy. Orally-active preparations of steroid hormones, especially oestrogens when taken post-coitally as an emergency measure, may cause delayed or arrested transport of the embryo in the Fallopian tube, thereby promoting the possibility of an ectopic implantation (see Coutinho 1971). As has been noted in previous chapters, normal functioning of the tubal musculature and cilia is extremely sensitive to the balance of circulating gonadal steroid hormones, and the post-coital treatment undoubtedly leads to disruption in these structures. There is evidence also of an association between the use of intra-uterine devices and the relative incidence of ectopic pregnancies. It is possible that the device occasionally acts to repel embryos from the cavity of the uterus towards or into the intra-mural segment; a changed pattern of myometrial contractions may be involved, and these in turn may be transmitted to the myosalpinx. The precise shape of the intra-uterine device and its location in the uterus at the time of ovulation could thereby have an important bearing on egg transport through the tubal lumen.

One other route to ectopic pregnancy can be mentioned here. Steptoe and Edwards (1976) first reported a tubal pregnancy after in vitro fertilization and transcervical transplantation of an embryo into the uterus and, with the increasing use of this approach to alleviating infertility, several other groups have since noted an elevated frequency of ectopic pregnancy after non-surgical transfer into the uterus. This is all the more remarkable since a deliberate attempt to instil an embryo into the Fallopian tube by the cervical approach would be deemed exceedingly difficult. Once again, an explanation for this experimental finding may be associated with an induced contractile activity in the myometrium and myosalpinx. But consideration should also be given to the abnormal state of the Fallopian tubes in women receiving such transfers: the absence of a full range of physiological functions, including cilial beat and fluid flow, may have rendered the tubal epithelium as receptive — to the transplanted embryo — as that of the uterus. Moreover, there may be some dilatation of the isthmus and intra-mural segment if secretions cannot escape into the peritoneal cavity.

Advice as to management of tubal pregnancies is, of course, beyond the scope of this text, but the overall objectives must clearly be (1) to remove the potential risks to the patient, especially those associated with rupture and haemorrhage, (2) to conserve as much of the tube as possible, and (3) to avoid compromising the function of the adjoining ovary, by striving to prevent the development of adhesions. However, with points (2) and (3) in mind, it is worth noting that most forms of tubal surgery are generally accepted to increase the risk of ectopic (or repeated ectopic) pregnancy. Nonetheless, conservation of as much of the tube as possible is receiving increasing emphasis in the North American literature (McComb 1986).

Many previous discussions as to why ectopic pregnancies should be peculiar to primates have focussed on (1) the invasive mode of implantation, with some emphasis on the properties of the syncytiotrophoblast and (2) the nature of the tubal epithelium. Because of the simplex uterus and the adjoining intra-mural portion of the tube, a continuity of epithelial types has been inferred. Furthermore, the absence of a specific utero-tubal junction would enable admixture of tubal and uterine secretions. This lack of a clear distinction between the uterine and tubal environments was frequently held to facilitate the initiation of tubal pregnancies. Yet other views have concentrated on subtle differences in the timing of egg transport in humans, especially on the relatively prolonged stay in the ampullary portion compared with egg passage in farm and laboratory animals (see Chap. IV; Croxatto et al. 1978). And finally, in women, if not sub-human primates, there has been consideration of emotional disturbances and the possibility of tubal spasm arresting passage of the embryo en route to the uterus.

In terms of physiological explanations for the existence of ectopic pregnancies, by far the most attractive one concerns modes of regulation of the lifespan of the corpus luteum. Whereas in farm and laboratory animals, hysterectomy leads to prolonged maintenance of the corpus luteum or corpora lutea (Anderson, Bland and Melampy 1969; Anderson 1973), due to removal of the tissues that elaborate the luteolytic hormone prostaglandin F_{2a}, the cyclic lifespan of the primate corpus luteum is not compromised by this operation. The current understanding is that ovarian factors rather than a uterine luteolysin are the principal initiators of regression of the cyclic corpus luteum, at least below the level of the pituitary gland and withdrawal of its trophic support. Because of the luteolytic potential of the uterus in farm and laboratory species, it is necessary that the embryos enter the uterus in order to exert an anti-luteolytic influence and thereby permit maintenance of the ovarian secretion of progesterone. By contrast, since this requirement seemingly does not exist in primates, development of the embryo in an ectopic site is possible in the sense that the non-gravid uterus will not be acting to promote luteal regression. A hypothesis to this effect was presented by Levasseur (1983).

Numerous studies have been performed in an attempt to develop an experimental model for ectopic pregnancies in laboratory and farm animals. These have been repeatedly unsuccessful, in part due no doubt to the reasons presented above, although it is worth noting that mouse trophoblastic tissue can proliferate in extra-uterine sites (Kirby 1965). The techniques for attempting to initiate ectopic pregnancies have usually involved placing ligatures around the Fallopian tubes at differing intervals after fertilization to prevent passage of embryos to the uterus. Development seldom proceeds beyond the early blastocyst stage (e.g. Murray et al. 1971; Pope and Day 1972), and diverse reasons could be invoked for failure of the embryo at this stage. It is not clear to this author, however, whether such models have been appropriately tested by means of transplantation of a separate set of embryos into the uterus of the same animal to prevent any precocious programming of luteolytic activity. At the very least, this approach would be worth testing and/or that of hysterectomy soon after mating of a polytocous animal, whilst leaving the Fallopian tubes and ovaries in situ. Development beyond the blastocyst stage might thereby be obtained.

Intra-Uterine Fertilization and Estes' Operation

Although fertilization of mammalian eggs normally takes place in the ampullary region of the Fallopian tube, various studies have examined the possibility of obtaining fertilization at sites in the reproductive tract other than the ampulla. Perhaps the earliest experimental evidence along these lines came from clinical studies in patients with obstructed Fallopian tubes. In these, a small number of pregnancies was reported following transplantation of the ovary into the uterus − Estes' operation (Fig. VIII.3) − and subsequent bouts of coitus. Reviews of such operations were presented by Estes (1909), Estes and Heitmeyer (1934), Preston (1953) and von Ikle (1961), with the overall incidence of pregnancy being approximately 10% (Table VIII.1). A more detailed quantitative analysis of suc-

Table VIII.1. A geographical survey of the success of Estes' operation as determined by the incidence of established pregnancy. (Modified from von Ikle 1961; Adams 1979)

Hospital location	No. of cases	No. becoming pregnant
New York	1	1
New York	50	4
Königsberg	9	0
Sofia	20	4
Rio de Janeiro	7	1
Stockholm	23	4
Nairobi	10	3
Groningen	16	0
Helsinki	128	8
Tarbes	5	3
St. Gallen	1	1
Total	270	29

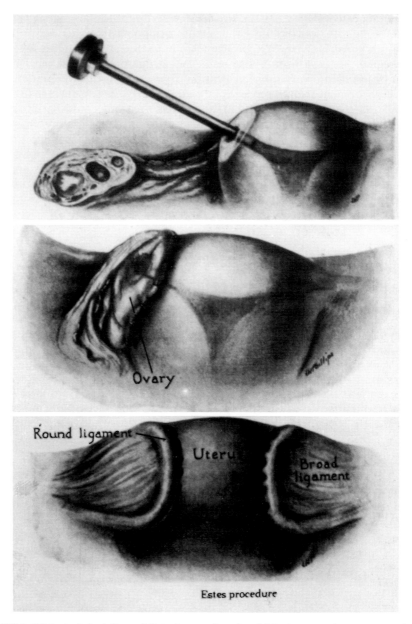

Fig. VIII.3. Historical depiction of Estes' operation, in which the ovary is auto-transplanted into the uterus in instances of obstructed Fallopian tubes

146

cess rates following Estes' operation has been given by Adams (1979). But it is uncertain from this literature whether both tubes were obstructed and completely removed (bilateral salpingectomy) in every case or whether some pregnancies might have resulted from a tubal fertilization after the surgical manipulations associated with the operation (Hunter 1968). That a proportion of recently-ovulated human oocytes can be fertilized in the uterus and undergo normal development is no longer in doubt (see below and Chap. IX), although the success of Estes' operation does raise other important questions, not least concerning the process of ovulation.

As an approach to studying further the low incidence of pregnancy in humans following Estes' operation, Heitmeyer (1934) transplanted rabbit ovaries into the uterus of the same animal; he was successful in obtaining ovulation in 12 rabbits, as demonstrated by both the presence of corpora lutea and recovery of eggs from the uterine lumen but, despite copulation of these animals with proven males, no pregnancies resulted. Similar negative results had previously been obtained in guinea pigs (Mairano and Placeo 1928), although the continuity of oestrous cycles inferred that the grafts were functional. These findings do not preclude the possibility of successful fertilization with a subsequent degeneration of eggs in the uterus. The deleterious influence of asynchrony in rabbits between the stage of embryonic development and that of the endometrium has since become well known (Chang 1950; Adams 1979). As a different experimental approach from that of Heitmeyer, Chang (1955) reported that transfer of 26 recently-ovulated rabbit eggs into the uterus of a mated recipient resulted in fertilization of 12 of these eggs as determined at recovery some 6 h later. Bedford (1969) pursued this approach of transplanting freshly-ovulated eggs into the uterus of previously-mated does primarily to determine just how soon ejaculated spermatozoa might achieve capacitation in the uterus. Eggs were recovered and examined towards the end of their fertilizable life 6−7 h after transfer. Thirty-four eggs from seven experiments were fertilized (94%) when the recipient does were mated some 12−13 h before transfer. Hence, exposure of spermatozoa to the Fallopian tubes was not essential for the completion of capacitation. Subsequent development of eggs could not, of course, be commented upon but degeneration rather than formation of blastocysts would have seemed inevitable.

In the golden hamster, an extensive series of experiments failed to obtain fertilization in the uterus. Recently-ovulated eggs in cumulus were transplanted from the tubes of 66 donors into the uterine horns of a similar number of mated or inseminated recipients at known stages of the cycle. None of 392 eggs was fertilized after periods of 2−8 h in the recipient uterus, nor were spermatozoa attached to or within the zona pellucida. Signs of degeneration were evident, especially the formation of deutoplasmic globules in the perivitelline space (Fig. VIII.4); this suggested an adverse influence of the tonicity of the uterine fluids upon the egg membranes (Hunter 1968), as previously postulated for the rabbit (Chang 1955). The possibility of failure to complete sperm capacitation in the hamster uterus was raised by this work, and a related study emphasized the synergism between uterus and Fallopian tubes in expediting this process in hamster spermatozoa (Hunter 1969). However, in an even larger series of transfers in the golden hamster, a small proportion of fertilized eggs has been obtained in the uterus (Cuasnicu; personal communication).

147

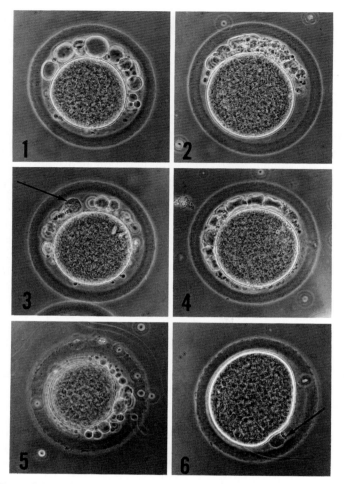

Fig. VIII.4. Eggs of the golden hamster recovered from the uterine lumen, into which they had been transplanted in an unsuccessful attempt to obtain fertilization in that site. Note the extensive formation of deutoplasmic globules in the perivitelline space (Hunter 1968)

Attempts to fertilize pig eggs in the uterus (another species in which the ejaculate is propelled through the cervix into the cornua) were reported by Baker and Polge (1973). Pre-ovulatory oocytes removed from follicles shortly before the anticipated time of gonadotrophin-induced ovulation were used in part of the study, whilst recently-ovulated eggs in cumulus were transplanted in another series. Only a small proportion (~5%) of such follicular and ovulated oocytes was fertilized in the uterus, and a possible contribution of the Fallopian tubes to sperm capacitation could not be ruled out in these instances. Sperm penetration of eggs in utero was not obtained when the tubes were ligated at their junction with the uterus before transplanting eggs to the latter site (Baker and Polge 1973). Again, perhaps of relevance here is the reported synergism between uterine and tubal environments in promoting capacitation of boar spermatozoa (Hunter and Hall 1974).

In stark contrast to the findings in experimental animals, recent studies in women have shown that secondary oocytes recovered just before ovulation can be introduced into the uterus with a suspension of ejaculated spermatozoa and give rise not only to fertilization but also to full term pregnancies (e.g. Craft et al. 1982). Because this approach avoids the steps of in vitro fertilization and culture of the embryo, it presents a number of technical advantages, although its overall incidence of success remains to be evaluated in diverse laboratories. The research breakthrough which pointed the way to these clinical achievements was the work of Marston and colleagues using rhesus monkeys. In a series of transplant studies, Marston, Penn and Sivelle (1977) questioned the dogma based on results from laboratory rodents and farm animals that primate embryos would also require a close synchrony between their stage of development and that of the endometrium for successful reception and growth. Marston further showed that young pronucleate eggs could be transplanted into the uterus of the same animal and give rise to viable pregnancies (Table VIII.2). As a group, therefore, primates may not require a strict synchrony between embryonic and uterine development, although overall pregnancy rates would doubtless be enhanced by embryos entering the uterus on a physiological timescale.

Table VIII.2. The results of transplanting embryos of the rhesus monkey from the Fallopian tube to the uterus. (Data courtesy of Dr. J. H. Marston)

Stage of embryonic development	Number of transplants	Number of successful pregnancies
Pronucleate	4	2
2-cell	3	3
4-cell	3	0
5 – 6 cell	11	4
7 – 8 cell	8	1

Tubal Resection and Reconstructive Surgery

The objectives of this section are, once again, not to cover clinical procedures in detail but rather to discuss experimental studies in domestic animals and to comment on their relevance to primates. Some of these experimental studies have been critically and concisely reviewed by Pauerstein (1978) and Pauerstein and Eddy (1979).

The aim of our own work involving tubal surgery was to clarify the rôle of the isthmus in the events of fertilization, there already being circumstantial evidence that this portion of the tube influences the number of spermatozoa progressing to the ampulla. Operating on both pigs and rabbits, most of the isthmus was resected from one of the tubes although a small proximal and distal stump

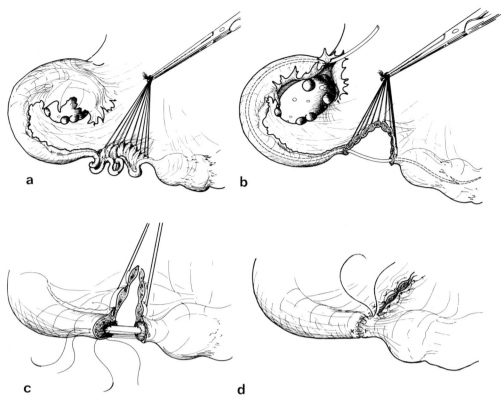

a   b

c   d

Fig. VIII.5. Tubal reconstructive surgery in the domestic pig to examine the influence upon the normality of fertilization of resecting most of the isthmus followed by end-to-end anastomosis. (Modified from Hunter and Léglise 1971 a)

remained in order to facilitate a satisfactory end-to-end anastomosis (Fig. VIII. 5). This procedure was performed with the support of a Teflon or silastic splint which was subsequently removed without recourse to further surgery (Hunter and Léglise 1971 a, b).

The striking finding from studies in both species was the significant increase in the incidence of polyspermic fertilization: 32% of the eggs recovered from the resected and anastomosed tubes of ten pigs showing this condition with usually two to three accessory male pronuclei clearly distinguishable in the vitellus. The eggs were prepared as histological sections, leaving no doubt as to the correct diagnosis of nuclear structures nor, with hindsight, should the finding of polyspermy have occasioned surprise. The voluminous ejaculate of the boar bathes the extremity of the Fallopian tubes by the completion of mating, usually with a suspension of spermatozoa containing $\geqslant 10^8$ cells per ml, and the utero-tubal junction and isthmus serve to regulate the number of spermatozoa passing to the site of fertilization. That polyspermy (7%) should have occurred at all in the rabbit as a sequel to resection of the isthmus was not anticipated, for the cervix and uterus are interposed between the site of ejaculation and that of fertilization. Nonetheless, the isthmus must contribute to the sperm gradient (Chap. IV), so the question arises from these findings as to whether a proportion of the failures to establish pregnancy in women after tubal reconstructive surgery might have been associated with abnormal fertilization (Hunter and Léglise 1971 a, b).

A subsequent study in pigs examined foetal survival and presented evidence for reasonable fertility after microsurgical removal of the isthmus (Paterson et al. 1981); in fact, the findings in this study and the earlier one described above are in close agreement. Reasonable fertility would still have been anticipated in our own experiments if the animals had been permitted to go to term for, whilst 32% of the eggs from the resected tube were polyspermic, 68% were normally fertilized, as were the eggs from the control tube. Such results would therefore parallel the 60% foetal survival reported by Paterson et al. (1981).

Surgery has also been performed in experimental animals to determine the influence, for example, of (1) removing the fimbrial extremity, on the effectiveness of egg capture at ovulation, and (2) reversing portions of the tube on the normality of egg transport, and thus to comment on the rôle of the cilia and myosalpinx. In the former instance, eggs can enter the tube and be fertilized (Metz and Mastroianni 1979; Beyth and Winston 1981), whilst in women with appropriate surgical modifications, eggs from one ovary may be captured by the contralateral tube, the so-called phenomenon of trans-abdominal migration (First 1954; Edwards 1980). Clearly, the egg does not migrate in an active sense but rather is drawn by currents of fluid thought to stem from the action of cilia. In this situation, the incidence of egg 'pick-up' from the pouch of Douglas is unknown.

Surgical modification at the other extremity of the tube, in the form of complete resection of the utero-tubal junction in rabbits, may compromise fertility (David, Brackett and Garcia 1969). However, further studies in rabbits have suggested that the utero-tubal junction is not critical to reproduction (Pauerstein 1978). The latter is also true of the ampullary-isthmic junction, for its removal in the rabbit does not interfere with normal fertility (Eddy, Antonini and Pauerstein 1977).

The functional involvement of specific regions of the Fallopian tube can be examined by surgical reversal rather than by actual removal. As far as the isthmus is concerned, reversal of 1-cm segments failed to interfere with fertility in rabbits (Eddy, Hoffman and Pauerstein 1976). By contrast, reversal of ampullary segments of similar length prevented pregnancy, seemingly due to the relatively dense array of cilia in the reversed portion beating the wrong way (i.e. in the original sense), thereby preventing an orderly progression of eggs to the site of fertilization at the ampullary-isthmic junction; eggs in cumulus are arrested at the interface with the reversed segment (Pauerstein 1978).

Evidence from human tubal surgery indicates that even when substantial portions of the isthmus are missing, re-anastomosis of the remaining tissues of the tube can lead to successful intra-uterine pregnancies. The need here may be primarily a question of mobility of the tube and its associated mesenteries at the time of ovulation. Hence, no absolute requirements for specific gamete or embryo interactions with this portion of the duct appear to exist, a conclusion that can be drawn from other lines of experimental evidence (see below). Likewise, after

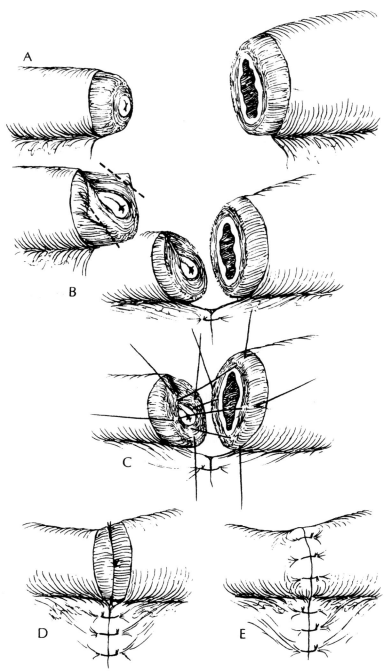

Fig. VIII.7. Two approaches to microsurgical reconstruction of the human Fallopian tube. Some principal steps in re-anastomosis of ampullary and isthmic portions with and without the support of a splint. (Adapted in part from Gomel 1977, 1978)

154

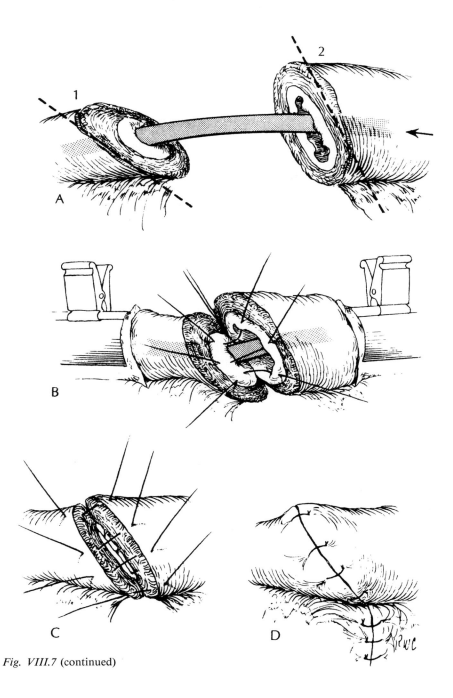

Fig. VIII.7 (continued)

155

Descriptions of clinical procedures can be sought elsewhere, but strict adherence to the principles listed below is a pre-requisite of success in microsurgery. The techniques aim to minimize trauma and injury to the delicate tissues surrounding and within the Fallopian tubes; avoiding haemorrhage in the mucosa is essential to restoration of a patent duct (Fig. VIII.7).

The main principles are:
1. A minimum handling of tissues, preferably with flame-polished glass rods.
2. Prevention of tissue dehydration by means of careful irrigation with isotonic fluid.
3. Meticulous preparation of the portions to be anastomosed, with unequivocal removal of all diseased tissue.
4. Exact alignment of the opposing ends of the tube.
5. Precise haemostasis of both the tube and the mesosalpinx.
6. Use of extremely fine, non-reactive suture material.
7. Placement of supporting sutures in the mesosalpinx to prevent tension in the anastomosed tissues.
8. Avoiding residual areas of raw tissue and subsequent adhesions by careful reperitonization.

Observations on the incidence of full-term pregnancies following such procedures have been presented by Winston (1980).

As already noted in this chapter, an alternative approach to problems of tubal infertility is that of laparoscopic oocyte recovery together with transplantation via the cervix; fertilization is attempted either in vitro or in vivo. Because the success of the laparoscopy and in vitro fertilization approach has gradually increased since the 1970s (see Chap. IX), and since these procedures avoid laparotomy, this approach must be seriously considered before opting for microsurgery. Indeed, some would argue that oocyte recovery and transfer now represents the method of choice. However, caution is advised before reaching a firm conclusion, especially if the incidence of ectopic pregnancies after trans-cervical egg transfer is of the order of 1 in 10, as some clinics are beginning to observe.

REFERENCES

Adams CE (1977) Ectopic pregnancy. Bibliog Reprod 30:97–98
Adams CE (1979) Consequences of accelerated ovum transport, including a re-evaluation of Estes' operation. J Reprod Fertil 55:239–246
Anderson LL (1973) Effects of hysterectomy and other factors on luteal function. In: Greep RO, Astwood EB (eds) Handbook of physiology, endocrinology II, Part 1. American Physiological Society, Washington, pp 69–86
Anderson LL, Bland KP, Melampy RM (1969) Comparative aspects of uterine-luteal relationships. Recent Progr Horm Res 25:57–99
Baker RD, Polge C (1973) Sperm penetration of pig eggs in utero. J Reprod Fertil 33:347–350
Bedford JM (1969) Limitations of the uterus in the development of the fertilizing ability (capacitation) of spermatozoa. J Reprod Fertil Suppl 8:19–26
Benirschke K (1969) Pathologic processes of the oviduct. In: Hafez ESE, Blandau RJ (eds) The mammalian oviduct. University of Chicago Press, pp 271–307

Beyth Y, Winston RML (1981) Ovum capture and fertility following microsurgical fimbriectomy in the rabbit. Fertil Steril 35:464–466

Brosens I, Winston R (1978) Reversibility of female sterilization. Academic Press, London

Brunham RC, MacLean IW, Binns B, Peeling RW (1985) *Chlamydia trachomatis*: its rôle in tubal infertility. J Infect Dis 152:1275–1282

Chamberlain G, Winston R (1982) Tubal infertility: diagnosis and treatment. Blackwell Sci Publ, Oxford

Chang MC (1950) Development and fate of transferred rabbit ova or blastocyst in relation to the ovulatory time of recipients. J Exp Zool 114:197–216

Chang MC (1955) Developpement de la capacité fertilisatrice des spermatozoides du lapin à l'interieure du tractus genital femelle et fécondabilité des oeufs de lapine. In: La fonction tubaire et ses troubles. Masson, Paris, pp 40–52

Coutinho EM (1971) Tubal and uterine motility. In: Diczfaluzy E, Borell U (eds) Control of human fertility. Nobel Symposium 15, Wiley, New York, pp 97–115

Craft I, McLeod F, Green S, Djahanbakhch O, Bernard A, Twigg H (1982) Human pregnancy following oocyte and sperm transfer to the uterus. Lancet i:1031–1033

Croxatto HB, Ortiz ME, Diaz S, Hess R, Balmaceda J, Croxatto HD (1978) Studies on the duration of egg transport by the human oviduct. II. Ovum location at various intervals following luteinizing hormone peak. Am J Obstet Gynecol 132:629–634

David A, Brackett BG, Garcia CR (1969) Effects of microsurgical removal of the rabbit uterotubal junction. Fertil Steril 20:250–257

Eddy CA, Hoffman JJ, Pauerstein CJ (1976) Pregnancy following segmental ishmic reversal of the rabbit oviduct. Experientia 32:1194

Eddy CA, Antonini R Jr, Pauerstein CJ (1977) Fertility following microsurgical removal of the ampullary-isthmic junction in rabbits. Fertil Steril 28:1090–1093

Edwards RG (1980) Conception in the human female. Academic Press, London

Eschenbach DA (1985) Pelvic inflammatory disease. IPPF Med Bull 19, No 3:1–3

Estes WL (1909) A method of implanting ovarian tissue in order to maintain ovarian function. Pennsylv Medic J 13:610–613

Estes WL Jr, Heitmeyer PL (1934) Pregnancy following ovarian implantation. Am J Surg 24:563–580

Fergusson ILC (1982) Laparoscopic investigation of tubal infertility. In: Chamberlain G, Winston R (eds) Tubal infertility: diagnosis and treatment. Blackwell Sci Publ, Oxford, pp 30–46

First A (1954) Transperitoneal migration of ovum or spermatozoon. Obstet Gynecol 4:431–434

Gomel V (1977) Tubal reanastomosis by microsurgery. Fertil Steril 28:59–65

Gomel V (1978) Salpingostomy by microsurgery. Fertil Steril 29:380–387

Gomel V (1980) Microsurgical reversal of female sterilization: a reappraisal. Fertil Steril 33:587–597

Green-Armytage VB (1959) Recent advances in the surgery of infertility. J Obstet Gynaec Brit Emp 66:32–39

Heitmeyer PL (1934) Pregnancy following ovarian implantation: experimental investigation. Am J Surg 24:571–580

Henry-Suchet J, Loffredo V (1980) Chlamydiae and mycoplasma genital infection in salpingitis and tubal sterility. Lancet i:539

Hunter RHF (1968) Attempted fertilization of hamster eggs following transplantation into the uterus. J Exp Zool 168:511–516

Hunter RHF (1969) Capacitation in the golden hamster, with special reference to the influence of the uterine environment. J Reprod Fertil 20:223–237

Hunter RHF, Hall JP (1974) Capacitation of boar spermatozoa: synergism between uterine and tubal environments. J Exp Zool 188:203–214

Hunter RHF, Léglise PC (1971 a) Polyspermic fertilization following tubal surgery in pigs, with particular reference to the rôle of the isthmus. J Reprod Fertil 24:233–246

Hunter RHF, Léglise PC (1971 b) Tubal surgery in the rabbit: fertilization and polyspermy after resection of the isthmus. Am J Anat 132:45–52

Hunter RHF, Baker TG, Cook B (1982) Morphology, histology and steroid hormones of the gonads in intersex pigs. J Reprod Fertil 64:217–222

Hunter RHF, Cook B, Baker TG (1985) Intersexuality in five pigs, with particular reference to oestrous cycles, the ovotestis, steroid hormone secretion and potential fertility. J Endocrinol 106:233–242

Ikle FA von (1961) Schwangerschaft nach Implantation des Ovars in den Uterus. Gynaecologia 151:95–99

Josso N, Picard JY, Dacheux JL, Courot M (1979) Detection of anti-Müllerian activity in boar rete testis fluid. J Reprod Fertil 57:397–400

Kirby DRS (1965) The role of the uterus in the early stages of mouse development. In: Wolstenholme GEW, O'Connor M (eds) Preimplantation stages of pregnancy. Ciba Foundation Symposium, Churchill, London, pp 325–339

Laing AJ (1979) Fertility and infertility in domestic animals, 3rd edn. Baillière Tindall, London

Levasseur MC (1983) Causes possibles de la fréquence des gestations extra-uterines chez la femme et de leur rareté chez les mammifères domestiques. Contracept Fertil Sex 11:1207–1213

Mairano M, Placeo F (1928) Sul comportamento degli autotrapianti ovarici peduncolati nella cavita dell'utero. G Accad Med Torino 91:206–215

Marston JH, Penn R, Sivelle PC (1977) Successful autotransfer of tubal eggs in the rhesus monkey (*Macaca mulatta*). J Reprod Fertil 49:175–176

Mauriceau F (1675) Traité des maladies des femmes grosses. Paris

McComb P (1986) The determinants of successful surgery for proximal tubal disease. Fertil Steril 46:1002–1004

Metz KGP, Mastroianni L Jr (1979) Dispensability of fimbriae: ovum pick-up by tubal fistulas in the rabbit. Fertil Steril 32:329–334

Murray FA, Bazer FW, Rundell JW, Vincent CK, Wallace HD, Warnick AC (1971) Developmental failure of swine embryos restricted to the oviducal environment. J Reprod Fertil 24:445–448

Nalbandov AV (1964) Reproductive physiology, 2nd edn. Freeman, San Francisco

Novak ER, Woodruff JD (1967) Novak's gynecologic and obstetric pathology, 6th edn. Saunders, Philadelphia

Novak J, Rubin IC (1952) Anatomy and pathology of the Fallopian tubes. In: Walton JH (ed) Ciba Clinical Symposia 4, No 6, pp 179–199

Palmer R (1960) Salpingostomy – a critical study of 396 personal cases operated upon without polythene tubing. Proc R Soc Med 53:357–359

Palmer R (1978) Reversibility as a consideration in laparoscopic sterilization. J Reprod Med 21:57–58

Palmer R, Madelenat P, Mendels E (1979) Les problèmes de l'établissement de statistiques valables dans la chirurgie des stérilités tubo-péritonéales. In: Oviducte et fertilité. Colloque SNESF, Masson, Paris, pp 354–366

Paterson PJ (1978) Tubal microsurgery – a review. Austral NZ J Obstet Gynaecol 18:182–184

Paterson PJ, Downing B, Trounson AO, Cumming IA (1981) Fertility and tubal morphology after microsurgical removal of segments of the porcine Fallopian tube. Fertil Steril 35:209–213

Pauerstein CJ (1974) The Fallopian tube: a reappraisal. Lea and Febiger, Philadelphia

Pauerstein CJ (1978) From Fallopius to fantasy. Fert Steril 30:133–140

Pauerstein CJ, Eddy CA (1979) The role of the oviduct in reproduction; our knowledge and our ignorance. J Reprod Fertil 55:223–229

Pope CE, Day BN (1972) Development of pig embryos following restriction to the ampullar portion of the oviduct. J Reprod Fertil 31:135–138

Preston PG (1953) Transplantation of the ovary into the uterine cavity for the treatment of sterility in women. J Obstet Gynaecol Brit Emp 60:862–864

Randall S (1986) Ectopic pregnancy. IPPF Med Bull 20, No 6, pp 1–2

Shirodkar VN (1967) Still further experiences in tuboplasty. In: Westin B, Wiqvist N (eds) Fertility and sterility. Excerpta Medica Fdn, Amsterdam, pp 353–360

Stallworthy J (1948) Facts and fantasy in the study of female infertility. J Obstet Gynaecol Brit Emp 55:171–180

Steptoe PC, Edwards RG (1976) Reimplantation of a human embryo with subsequent tubal pregnancy. Lancet i:880–882

Sweet RL (1982) Chlamydial salpingitis and infertility. Fertil Steril 38:530—533

Tuffrey M, Falder P, Gale J, Quinn R, Taylor-Robinson D (1986) Infertility in mice infected genitally with a human strain of *Chlamydia trachomatis*. J Reprod Fertil 78:251—260

Vigier B, Tran D, Legeai L, Bézard J, Josso N (1984) Origin of anti-Müllerian hormone in bovine freemartin fetuses. J Reprod Fertil 70:473—479

Winston RML (1980) Microsurgery of the Fallopian tube: from fantasy to reality. Fertil Steril 34:521—530

Woodruff JD, Pauerstein CJ (1969) The Fallopian tube: structure, function, pathology and management. Williams Wilkins, Baltimore

In Vitro Fertilization, Manipulation of Eggs and Embryos, and Subsequent Transplantation

CONTENTS

Introduction 160
Historical Aspects 161
Experimental Conditions for In Vitro Fertilization 163
Clinical Approach to In Vitro Fertilization 165
Embryo Culture and Transplantation into the Uterus 168
Alternative Approaches to Infertility 171
Sperm Penetration of Zona-Free Oocytes 172
Other Cellular and Nuclear Manipulations 174
References 177

Introduction

The fertilization of mammalian eggs outside the body, that is penetration of the egg by a spermatozoon in some form of cell culture system, for many years remained a research target for physiologists and scientists in related disciplines. Because mammalian fertilization normally occurs within the confines of the Fallopian tubes, one much repeated argument has been that a system of in vitro fertilization would be a valuable analytical tool for examining biochemical and physiological factors that contribute to the successful union of gametes. Furthermore, in the context of a technology that might one day be applied in the field of animal breeding, in vitro fertilization has also been suggested as a possible sequel to in vitro maturation of large numbers of oocytes and as a means of fertilizing eggs obtained from immature animals or after excessive responses to superovulatory treatments. And of course, in the case of human medicine, a technique of in vitro fertilization is already being used to overcome problems of infertility due to blocked Fallopian tubes. There is no question here of the emotive testtube baby, since the embryo is instilled into the uterus of the mother at an early stage of cleavage (4–8-cell). The subject of in vitro fertilization has been reviewed on numerous occasions, but reference to the works of Austin (1961a), Chang (1968), Thibault (1969), Gwatkin (1977), Edwards (1980), Biggers (1981), Brackett (1981), Wright and Bondioli (1981) and Hartmann (1983) may be found helpful at the outset.

Because in vitro fertilization must first depend on obtaining suitable preparations of eggs and spermatozoa, it should be clear that fertilization itself will represent only one step in a sequence of manipulations, and that in order for an embryo formed in vitro to develop to term, some means of transplantation will have to be applied. The paragraphs that follow will not attempt to treat the subject exhaustively, but will present many of the key details involved in bringing male and female gametes together under appropriate conditions of culture. Alternative approaches to solving problems of infertility will also be discussed.

Historical Aspects

The history of attempted in vitro fertilization in mammals is a lengthy one, dating back at least to the work of Schenk (1878) with rabbit and guinea-pig eggs; many of the well-documented attempts are described by Austin (1961 b) and Thibault (1969). It is generally believed, however, that before the requirement for capacitation of spermatozoa was appreciated, presumptive fertilization of eggs in vitro with samples of ejaculated spermatozoa or those prepared from the male tract was either a chance occurrence or, more probably, an incorrect diagnosis. In the latter case, the most common pitfall was to regard eggs containing two pronuclei or other nuclear structures as penetrated and fertilized, whereas such a situation could have arisen by parthenogenetic activation, especially under less than perfect conditions of culture. The essential criteria for establishing fertilization in a recently penetrated egg should include the presence of first and second polar bodies, male and female pronuclear elements, and detection of some remnant of the sperm mid-piece or tail. Other misleading observations have been those on 2- and 4-cell eggs that were in fact fragmenting, and the appearance of unswollen sperm heads seemingly within the cytoplasm of histologically-sectioned eggs when it is now clear that they must have been displaced by the microtome. Yet a further source of error, albeit leading to eventual fertilization, has been the transplantation of eggs to the Fallopian tubes of recipients soon after attempted fertilization in vitro in order to examine their development. The obvious danger here is of transferring spermatozoa in association with unfertilized eggs which then undergo successful penetration and cleavage in the recipient. Paradoxically, this very situation may now be used to overcome certain instances of human infertility (see below).

Having stressed the more frequent sources of error in interpreting putative claims for in vitro fertilization, and inferred the need for caution in evaluating contemporary attempts, the first authentic successes and state of the art today can now be considered. Capacitation was recognized as a specific phenomenon by Austin (1951) and Chang (1951) (see Chap. V), and the first systematic studies of in vitro fertilization using spermatozoa recovered from the female tract (i.e. spermatozoa presumed to have been capacitated in vivo) were almost certainly those by Dauzier, Thibault and Wintenberger (1954) in the rabbit. In addition to the use of spermatozoa flushed from the tubes and uterus some 12 h after mating, these authors employed an incubator system with rotating glass tubes in which the recently-ovulated eggs were agitated for 30–120 min in a small volume of Locke's solution before introducing the sperm suspension. This washing treatment was held to improve the proportion of eggs fertilized due to removal of a fertilizin-like substance from the egg or its investments (Thibault and Dauzier 1960, 1961). Although Thibault and his colleagues presented impeccable histological evidence of in vitro fertilization as shown in pronuclear and cleaved eggs, it was Chang (1959) who consummated the experimental procedure of the French workers by transplanting rabbit eggs after fertilization and cleavage in vitro to suitable recipients, in due course obtaining viable young. But these early achievements with the rabbit egg failed to point the way to a fruitful in vitro technique with the gametes of large domestic animals, even though Thibault and Dauzier (1961) recorded incidental successes for the sheep and pig.

Whilst these first studies and much later work obtained sperm samples from the tract of mated females, more recent experiments – especially in rodents but also in humans – have obtained capacitation of epididymal or ejaculated spermatozoa in vitro. The forerunners here were Yanagimachi and Chang (1963, 1964) who obtained in vitro fertilization of golden hamster eggs with suspensions of epididymal spermatozoa, this step forward being followed by fertilization in vitro of hamster ovarian oocytes using preparations of epididymal spermatozoa (Barros and Austin 1967). Thus, capacitation of hamster spermatozoa together with any final maturation of the oocyte must have been obtained in the culture system, neither gamete being exposed to the fluid environment of the female tract. As far as spermatozoa of the golden hamster are concerned, Bavister (1969) has put forward the view that capacitation in this species may simply require provision of a suitable – albeit critically defined – fluid milieu in which this intrinsically regulated process will occur. Use of oocytes aspirated from follicles shortly before or at the moment of ovulation has proved valuable in the rabbit (Seitz, Brackett and Mastroianni 1970) and human (Edwards, Steptoe and Purdy 1970), and

Table IX.1. Examples of species in which in vitro fertilization has been obtained. With the notable exception of the golden hamster, mouse and man, the results of individual attempts have been extremely variable and unpredictable. (Modified from Hunter 1980)

Species	Two or more reports of in vitro fertiliza- tion[a]	Foetuses and/or viable offspring produced after transplantation
Rabbit	+	+
Rat	+	+
Mouse	+	+
Golden hamster	+	−
Chinese hamster	+	−
Guinea pig	+	−
Cow	+	+
Sheep	+	+
Pig	+	+
Man	+	+

+ indicates positive evidence; − indicates not determined.

[a] Sperm penetration of oocytes alone has not been considered sufficient evidence of fertilization, as this condition may be found in immature or atretic eggs. The criteria insisted upon include activation of the second meiotic division with extrusion of the second polar body and formation of a female pronucleus together with development of a male pronucleus, or stages subsequent to these.

possibly gave impetus to subsequent experiments on in vitro fertilization of in vitro matured oocytes.

Species in which in vitro fertilization has been obtained in at least several trials are listed in Table IX.1. In the pig, von Harms and Smidt (1970) recorded some apparent fertilizations using ejaculated boar spermatozoa, and more recently Iritani, Niwa and Imai (1978) have presented photographic evidence with which to support their own observations. In vitro capacitation of boar spermatozoa was improved by raising the culture temperature to 39 °C, permitting subsequent in vitro fertilization (Cheng 1985). In vitro fertilization of sheep eggs has been reported by several groups, and the same is true of bovine oocytes (e.g. Ball et al. 1983; Iritani et al. 1984). The birth of a calf as a sequel to such procedures was first reported by Brackett et al. (1982). An impressive fact in reviewing much of the data on in vitro fertilization is the enormous variability in the proportion of eggs fertilized between experiments in any given species, the only exceptions to date being the golden hamster and possibly the mouse, which may regularly yield an incidence of fertilization of 80–90% or more. Why should this variability exist? Apart from differences in skills between individuals, and the possible advantage of working with certain strains of animal whose gametes may be particularly amenable to manipulations in vitro, the actual conditions in the culture system play a vital rôle.

Experimental Conditions for In Vitro Fertilization

As judged from extensive experimentation in laboratory animals, careful attention to the following factors is an essential prerequisite of success; the list is by no means exhaustive.

1. The source and condition of spermatozoa. Suspensions of spermatozoa recovered from the female tract several hours after mating featured in many experiments during the 1960s, but washed ejaculated or preincubated epididymal spermatozoa are nowadays used extensively in rodent and primate studies. The process of liquefaction must first be awaited in the human ejaculate, requiring a period of about 30 min (Mann 1964). A washing treatment usually implies mild centrifugation (300 g) to remove most of the free seminal plasma, followed by a resuspension of the sperm cells in the culture medium. This is best achieved by the swim-up technique, in which highly motile spermatozoa migrate out of the centrifuge pellet into the medium above.

2. The presence or absence of follicular cells and fluid around the eggs at the commencement of incubation. The majority of studies, especially in rodents, have used recently-ovulated eggs recovered within a plug of cumulus cells by perfusion or rupture of the Fallopian tubes.

3. The choice of a medium, physiological salt solutions such as Locke's, Tyrode's or Krebs-Ringer being at first favoured. Recent studies have used more complex media such as TC 199, Ham's F-10 or Dulbecco's phosphate-buffered saline with HEPES. An osmolality of 290–310 mOsmols and a pH of 7.4–7.6 are common, the latter frequently being maintained with a bicarbonate buffer and

163

a gas phase of 5% CO_2 in humidified air. In addition to follicular or tubal fluid constituents, the medium usually contains a large molecular weight substance such as serum albumin (concentration of $1-5$ g/l) or foetal cord serum or a synthetic substance such as polyvinyl pyrrolidone. In studies during the last decade, the presence of lactate and calcium ions in the culture medium has been considered desirable or essential.

4. The volume of the microdrops used for culture and the concentration of the sperm suspension therein, respective figures often being of the order of $40-100$ µl and $1-2\times10^6$ cells/ml. Although such sperm numbers would greatly exceed those reaching the ampulla at the time of fertilization, it is generally accepted that only a small proportion of the spermatozoa would be competent to penetrate the eggs (see Chaps. IV and V).

5. The time relationship between placing the separate gamete preparations in culture, and then mixing the two. A period of preincubation of the sperm suspension, perhaps in a medium rich in ovarian follicular fluid, is usually found to be beneficial.

6. The absolute minimization of temperature fluctuations during the in vitro manipulations; the avoidance of contamination of the cell suspensions with blood which may rapidly cause agglutination of the spermatozoa; and the strict observance of aseptic precautions. The culture system is normally held in darkness, except during periodic examination under the dissecting microscope.

As to the actual system of culture, microdrops of medium in Falcon plastic Petri dishes have been most frequently used, with the advantage that the cell preparations can be inspected under the microscope without disturbance. But as

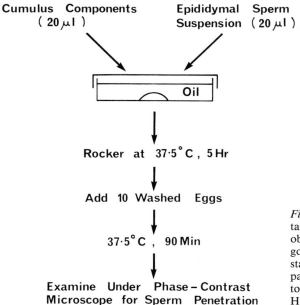

Fig. IX.1. Diagrammatic representation of a simplified sequence for obtaining in vitro fertilization of golden hamster oocytes. The first stages in the sequence permit capacitation of epididymal spermatozoa. (After Gwatkin 1977; Hunter 1980)

already mentioned, rotating glass tubes containing the medium between droplets of paraffin oil have been used by Thibault and Dauzier (1961), and others have used glass or plastic micro test-tubes, watch glasses or cavity slides. To avoid problems of evaporation, a common practice with the Falcon dishes has been to place the microdrops of medium beneath a thin layer of paraffin or mineral oil, the oil having been previously equilibrated against a small volume of the culture medium to avoid subsequent osmotic disturbances in the microdrops. A laboratory sequence for obtaining in vitro fertilization of hamster eggs is illustrated in outline in Fig. IX.1.

Clinical Approach to In Vitro Fertilization

As noted in the preface, studies on in vitro fertilization of human oocytes have expanded enormously since the report of a live birth from such a technique by Steptoe and Edwards (1978). This research activity has been reflected in an ever-increasing number of scientific meetings together with the publication of conference volumes and specialist journals. Some of the most recent books include those of Beier and Lindner (1983), Wood and Trounson (1984), Testart and Frydman (1985), Rolland et al. (1985), Jones et al. (1986) and Feichtinger and Kemeter (1987). Keeping abreast of all the latest experimental treatments and permutations is not easy, but the underlying principles can certainly be summarized. In fact, many of the most recent clinical approaches to infertility – such as intra-peritoneal or intra-uterine insemination – follow directly from studies in domestic animals, so the experimental ideas themselves are not new.

Important steps in a clinical approach to in vitro fertilization (Fig. IX.2) include the following:

– Stimulation of ovarian follicles.
– Monitoring of follicular maturity (i.e. size).
– Laparoscopic or ultrasonic-guided recovery of oocytes.
– Evaluation and maintenance of oocytes in culture.
– Preparation of the sperm suspension.
– Semination of an appropriate number of spermatozoa.
– Actual technique at the bench.

The classical approach to promoting growth of ovarian follicles is to inject either gonadotrophic hormones themselves or drugs such as clomiphene citrate that act to enhance release of endogenous gonadotrophins. Modern approaches to follicular stimulation that are proving valuable in farm animals, such as (a) modifying ovarian steroid feedback by passive immunization (Land et al. 1982), or (b) injecting or feeding drugs that interfere with steroid hormone biosynthesis or feedback (Webb 1987) or (c) injecting non-steroidal follicular fluid components such as inhibin to modify feedback (McNeilly 1985; Wallace and McNeilly 1986) appear not yet to have undergone widespread clinical tests.

As to the timing of treatment, gonadotrophic hormones are best injected in the early follicular phase rather than in the presence of an active corpus luteum. A practical régime might be daily intra-muscular injections of 150 i.u. follicle

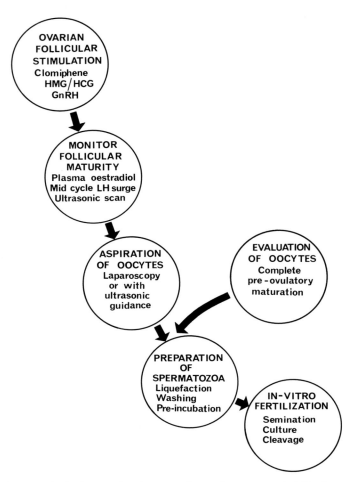

Fig. IX.2. Some principal steps in the procedure of recovery of follicular oocytes from stimulated ovaries and their subsequent fertilization in vitro. Although zygotes cultured to cleavage stages have usually been transferred, pregnancies have also been obtained by introduction of male and female gametes directly into the Fallopian tubes or uterus

stimulating hormone and luteinizing hormone (Pergonal; Serono) from Days 4 until 10 of the menstrual cycle or clomiphene stimulation from Days 5−9 of the cycle followed by 5000−10000 i.u. HCG intra-muscularly on Day 12, with laparoscopy and follicular aspiration scheduled 36 h later. The HCG injection acts to mimic the pre-ovulatory LH surge. Pulsatile intra-venous administration of gonadotrophin-releasing hormone (GnRH) or HMG by means of a perfusion pump has also been used, but this approach may seem less attractive to many patients and more suitable for cases of anovulatory infertility. On the other hand, because the pulsatile treatment with GnRH simulates hypothalamic secretion and induces release of endogenous gonadotrophins, it is more controlled and less likely to result in problems of ovarian hyperstimulation.

Growth of Graafian follicles was traditionally monitored by the titres of urinary oestrogens or serum oestradiol, or by detecting the onset of the mid-cycle pre-ovulatory gonadotrophin surge (i.e. LH), but physical estimates of follicular number and maturity are currently in favour because of their rapidity and greater accuracy. Ultrasonic scanning of the ovaries to delineate follicular growth has been referred to in Chapters IV and V. The objective of the technique is to determine the diameter of stimulated follicles: a measurement of 18 – 20 mm is considered appropriate for oocyte extraction. Expansion and progressive dispersal of the cumulus mass surrounding the oocyte has also been monitored by echo-sonography, this giving a more sensitive estimate of the time for oocyte aspiration (Bomsel-Helmreich 1985; Demoulin 1985). The aim is to recover oocytes after completion of the first meiotic division, at which stage the first polar body would have been fully extruded with the oocyte chromosomes arranged at second meiotic metaphase – just as at spontaneous ovulation.

The technique of laparoscopy has frequently been described in clinical texts (e.g. Steptoe 1977; Fergusson 1982; Steptoe and Webster 1982). Ultrasonic guidance of follicular puncture by the transvaginal route is now also well known (Wikland et al. 1983; Dellenbach et al. 1985). Whilst the aim is clearly to puncture the wall of the pre-ovulatory follicle and recover the oocyte within its cumulus cell investment, together with some follicular fluid, a requirement is also for the collapsed follicle to form a functional corpus luteum. Adequate secretion of progesterone is needed to prepare the uterine epithelium for the transplanted embryo(s), otherwise enhanced embryonic mortality would be anticipated (Wilmut, Sales and Ashworth 1986). Accordingly, a minimum of trauma should be caused by the aspiration needle, and any excessive aspiration of granulosa cells avoided. There continues to be debate as to whether defective luteal phases arise from procedures of follicular aspiration (Edwards et al. 1980; Garcia et al. 1981; Kerin et al. 1981; Frydman et al. 1982). Of some relevance may be the observation that the luteal phase is prone to endocrine disruption after hyperstimulation of human ovaries with pulsatile FSH, even without subsequent follicular aspiration (Messinis and Templeton 1987).

In the physiological situation, oocytes change fluid environments quite dramatically at ovulation, passing from the follicular antrum in which they have resided for many weeks or months into the ampulla of the Fallopian tube. In laboratory and farm animals, there is some evidence that this change of fluid environment corresponds to a phase of post-ovulatory maturation, at least as far as the oocyte surface and investing cells are concerned. But, apart from this putative phase of post-ovulatory maturation, oocytes obtained at laparoscopy will require a period of culture (e.g. 1 – 6 h) corresponding to the interval prior to spontaneous ovulation when they were removed from the follicle (see Trounson et al. 1982). They are maintained in droplets of medium under oil in Falcon plastic dishes before addition of the sperm suspension and their viability is assessed on morphological criteria. The physical conditions of culture have been extensively described in the volumes cited earlier in this section, the prime requirements of temperature control, pH and osmolarity being discussed above.

Reference has also been made above to sperm suspensions and their preparation in experimental animals. Whereas epididymal spermatozoa have commonly been used in laboratory species, the ejaculate would invariably be the source of spermatozoa used in human in vitro fertilization. Removal of seminal plasma by 'washing' and mild centrifugation is therefore a prerequisite, and this may be followed by a period of preincubation in a hypertonic solution to promote the process of capacitation. Evaluation of sperm quality concentrates on both morphology and progressive motility. Sperm concentration needs to be reduced in procedures for in vitro fertilization, as occurs in vivo during ascent of the female genital tract, otherwise the problem of polyspermic penetration will arise. The incidence of polyspermy in human eggs in vitro was reported to decrease from 5.5% at a concentration of 10×10^4 to 1.4% with 5×10^4 motile spermatozoa per ml. At $1-2.5 \times 10^4$ motile spermatozoa per ml there was zero polyspermy (Wolf et al. 1984). At the time of writing, the number of spermatozoa introduced into the droplets of fluid containing oocytes (i.e. seminated, to use the modern parlance) would be of the order of 50000 motile cells per ml in most laboratories. The number of functional or competent spermatozoa is the critical aspect, but the competence of a spermatozoon can only be fully evaluated by the process of fertilization itself. Any suspicion of male subfertility is compensated for by increasing the number of spermatozoa introduced in vitro.

The actual bench technique is invariably addition of the sperm suspension to the culture dish containing the oocyte in modified Ham's F-10 medium plus 7.5% heat-inactivated human foetal cord serum or peripheral blood serum from the patient. Individual oocytes are treated in separate dishes to facilitate microscopic inspection and subsequent procedures. However, in cases of oligospermy, when the number of cells available for semination may be severely restricted, oocytes are cultured together in a single dish. As to the incubation period for the gamete mixture, this is conventionally some $12-18$ h, which includes the overnight period. The oocytes are then inspected for penetration and pronuclear formation and transferred into fresh medium, thereby avoiding any deleterious influence of degenerating sperm products. Embryos judged to be viable are generally cultured for a further 24 h, so that they have embarked upon distinct mitotic divisions at the time of transplantation into the uterus.

Embryo Culture and Transplantation into the Uterus

The dogma from extensive studies in laboratory and farm animals is that survival of transplanted embryos depends on carefully timed procedures. Optimum results follow from transfer of embryos at a stage of cleavage matching the corresponding uterine environment. Thus, the requirement is for a so-called synchronization between stage of embryonic development and that of endometrial proliferation (Chang 1950). Underlying this physiological constraint are (1) the changing substrate needs for development of the unattached embryo, obtained from the luminal fluids of the tract (Chaps. III and IV), and (2) the dynamic nature of secretions which are programmed both quantitatively and qualitatively by the ratio of ovarian steroid hormones (Chap. III). Synchronization is best understood by recalling that, under normal conditions of fertilization, formation of the embryo and formation of the corpus luteum commence at almost the same time after completion of ovulation.

Successful transfer of pig embryos to the uterus requires them to be at the 4-cell stage of development or later, and the corresponding stage for cow and sheep embryos is 8 – 16-cells (Polge 1972; Betteridge 1977; Hunter 1980). Because human embryos enter the uterus from the Fallopian tubes at the 8 – 16-cell stage (Chaps. VI and VII), preliminary studies involving fertilization in vitro naturally attempted to culture embryos to such stages before introducing them to the uterus (Edwards, Steptoe and Purdy 1970; Steptoe and Edwards 1976). This approach offered the advantage that the morphological state of the cleaving embryo could be assessed under the dissecting microscope before transplantation. As noted in the previous chapter, however, it was becoming clear by the late 1970s that primate embryos did not suffer from the constraints of synchronization portrayed above for domestic species: pronucleate embryos could be transplanted to the uterus of recipient rhesus monkeys and establish successful pregnancies (Chap. VIII).

General remarks on the culture of embryos in vitro have been presented in Betteridge (1977), Daniel (1978) and Hunter (1980), together with specific comments on the culture of rodent and farm animal embryos (see also Wright and Bondioli 1981). Whilst there are many variations and, at present, individual preferences in the actual techniques of culture, most studies have used either Falcon plastic Petri dishes containing the embryos in droplets of medium under oil or glass tubes (mini test tubes) with silicone rubber or glass stoppers. Others have employed glass cavity slides, embryological watch glasses, or even slowly rotating glass tubes. An exceptional technique for short-term storage (10 h) has been the use of a length of silicone rubber capillary tubing (dimensions: 50 cm × 2.1 mm OD) by Baker and Dziuk (1970) to transport 2 – 8-cell pig embryos by air before successful transplantation into the uterus of a synchronized recipient. It is not clear from the report if a further cell division occurred in this in vitro situation in which the tubing was immersed in medium at 24 °C. General comments concerning media for culture, attention to pH and osmolarity, the use of equilibrated oil and other such details are, in principle, in line with those for in vitro fertilization. But use of a phosphate rather than bicarbonate based buffer has sometimes been favoured to avoid the necessity of the CO_2-enriched gas phase for maintaining pH. When blood serum is included as a component of a culture medium, it is essential that it be heat-treated (i.e. 56 °C for 30 min) to denature an ovicidal factor first described by Chang (1949).

In contrast to most reports for the embryos of domestic animals, human eggs fertilized in vitro seem to be able to continue development in culture at least to the morula or early blastocyst stage (Edwards, Steptoe and Purdy 1970), and some may develop further after retransfer to the uterus (Steptoe and Edwards 1976) or indeed give rise to a healthy baby (Steptoe and Edwards 1978). Thus, the notion of there being at least one sensitive stage in the early cleavage of mammalian embryos, as discussed for certain farm species (Hunter 1980), may not have a parallel in the primate.

The overall objective of transplantation is clearly to introduce the embryo aseptically into the uterus suspended in a minute volume of culture fluid with the minimum of disturbance to the maternal tissues. Details of the various techniques

Table IX.2. The increasing incidence of implantation with increasing number of human embryos introduced into the uterus following various means of ovarian stimulation. (Adapted from Edwards 1984)

Ovarian treatment	No. of embryos replaced		
	1	2	3
Clomiphene/LH surge	59/380	60/222	16/44 (36.4%)
Clomiphene/HCG	18/137	40/157	16/68 (23.9%)
Clomiphene/HMG/LH surge	5/52	10/39	15/43 (34.9%)
Clomiphene/HMG/HCG	6/25	10/41	42/117 (35.9%)
Total	88/594 (14.8%)	120/459 (26.1%)	89/272 (32.9%)

are available in many reviews (for domestic species: Polge 1972; Betteridge 1977, 1986; Hunter 1980; for human embryos, the numerous volumes cited above together with Steptoe, Edwards and Walters 1986). The point is worth making that no matter how delicate the instrumentation, all transplantation procedures will be invasive to a degree and thereby modify the physiological state of the genital tract. A long-standing observation from animal studies is that dilating the vagina with a speculum and broaching the cervix with a catheter or pipette invariably lead to a modified pattern of smooth muscle contractions. The contractions, in turn, may rapidly displace the embryo(s) from the original site of deposition. The result of such stimulated activity has been monitored in animals by means of radio-labelled artificial eggs (resin spheres) which may be ejected from the uterus back into the vagina within a few hours of transfer (Harper, Bennet and Rowson 1961; Rowson, Bennet and Harper 1964). A consequence of the manipulation in women could be the significant increase in tubal (ectopic) pregnancies (Chap. VIII).

As to the number of embryos introduced into the uterus (Table IX.2), it is now widely accepted from many human studies that the incidence of pregnancy in-

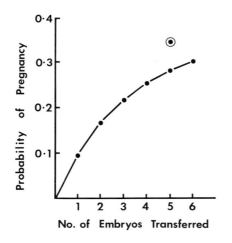

Fig. IX.3. Graphical representation of the probability of establishing human pregnancy as a function of the number of embryos transplanted. (Adapted from Trounson et al. 1985)

170

creases with the number of embryos transplanted (Fig. IX.3). On the other hand, consideration needs to be given to the obstetrical risks associated with multiple pregnancy and the high abortion rate, such as the 28% recorded in the comprehensive survey by Steptoe et al. (1986).

Alternative Approaches to Infertility

Whilst procedures of in vitro fertilization have captured the headlines, there are a number of other approaches to infertility which require brief mention; an underlying assumption in these approaches is that an ovary and at least one of the Fallopian tubes are functional.

The first is to transfer oocytes recovered at laparoscopy together with an appropriate sperm suspension directly into the ampulla of the tube via the fimbria − the so-called Gamete intra-Fallopian Transfer (GIFT). This procedure is also achieved by means of laparoscopy (Asch et al. 1984; Molloy et al. 1986). In one sense, at least, this concept is not new since surgical insemination of sperm suspensions into the lumen of the Fallopian tubes was performed nearly 40 years ago in laboratory animals (Austin 1951; Chang 1951). In farm species, unfertilized oocytes were already being transferred directly into the ampulla of inseminated animals by the mid 1960s (Hunter 1967; Hunter, Lawson and Rowson 1972), with fertilization occurring in many of the eggs. A notable observation with clinical use of the GIFT procedure − as with uterine transplantation of oocytes or embryos in women − is again that the incidence of pregnancy is dependent upon the number of eggs introduced into the ampulla. In the best hands and with a suitable selection of patients, some 30% or more may become pregnant as a sequel to the GIFT technique (Asch et al. 1984, 1985). Not only have gametes been transferred to the tubes, but also young embryos, i.e. zygotes, which may here be a less emotive term. The risk of ectopic and/or multiple pregnancies after the GIFT procedure still awaits a rigorous assessment.

A second approach that follows from experimental studies in animals is insemination of a sperm suspension directly into the peritoneal cavity; this has been shown to be a route to normal fertilization in the cow (Skjerven 1955), rabbit (Dauzier and Thibault 1956; Hadek 1958; Rowlands 1958), guinea-pig (Rowlands 1957) and domestic pig (Hunter 1978). Assuming a suitable quality of sperm suspension, the most critical factor influencing fertilization was found to be the timing of insemination relative to the moment of ovulation. Intra-peritoneal insemination was satisfactory in animals only when it was performed shortly before or at the time of ovulation, presumably because reservoirs of spermatozoa are not formed in the isthmus of the tube after this manipulation. However, the heightened activity of the tubal musculature, fimbria and associated mesenteries at the time of ovulation, together with the direction-oriented beat of the cilia (see Chap. IV), may also be vital for displacing spermatozoa into the tubal ostium.

In clinical studies, suspensions of spermatozoa washed free of seminal plasma have likewise been injected into the abdominal cavity close to the time of ovulation (35 h post-HCG injection after prior follicular stimulation). Imminent ovulation could be verified by ultrasonic scanning. The sperm suspension, diluted in

Ham's F-10 medium with 20% human albumin and prepared by the swim-up technique, was injected in 2 ml fluid under sterile conditions through the posterior vaginal fornix into the pouch of Douglas using a 19 gauge $\times$ 2.2 cm needle (Forrler et al. 1986a). As to sperm numbers, $0.6-18 \times 10^6$ cells were introduced and, apparently, at least 500000 motile spermatozoa had to be instilled if fertilization was to occur (Forrler et al. 1986b). This intra-peritoneal approach has been used in patients with periods of infertility ranging from $4-9$ years (mean duration 6.5 years), in which infertility was thought to be associated principally with the inability of spermatozoa to penetrate the cervical mucus. A pregnancy rate of 14% per treatment cycle has been reported in such instances (Forrler et al. 1986b).

A third form of treatment for problems of unresolved infertility, especially when poor semen or mucus quality is suspected, has been that of intra-uterine insemination. Again, this route had been previously employed in animal experiments and indeed − by means of trans-cervical passage of a Cassou straw − it is the most widespread method of artificial insemination in cattle. In patients with normal ovarian function but long-term infertility, intra-uterine insemination was performed when intra-cervical insemination had failed during a period of months (at least 6 months). In one series, ovarian stimulation involved injection of Pergonal: the intra-uterine insemination of spermatozoa washed free of seminal plasma followed $32-36$ h after an ovulating injection of HCG or when the dominant follicle measured more than 18 mm in diameter as determined ultrasonically. At least six pregnancies have been recorded after superovulation procedures, representing a rate of 18% per cycle (Serhal and Katz 1987).

Prior to this report, the results of several trials involving intra-uterine insemination had been negative (e.g. Irvine et al. 1986; Thomas et al. 1986), despite the promising results first presented by Kerin et al. (1984) and the limited success of Yovich and Matson (1986) with repeated insemination. An useful historical survey from 1957 is presented in the paper of Toffle et al. (1985), and a critical review of intra-uterine insemination has also been published (Allen et al. 1985). But, even accepting a degree of controversy over the value of so-called high (deep) intra-uterine insemination, especially in the absence of ovarian follicular stimulation, repeated attempts with this route may be warranted before resorting to the more invasive procedures of intra-peritoneal insemination (i.e. culdoscopic insemination) or GIFT or indeed in vitro fertilization. As a comment from studies in experimental animals with reduced numbers of spermatozoa, the timing of insemination may be more critical than previously realized if fully-capacitated human spermatozoa have a very short lifespan (see Hunter 1987).

Sperm Penetration of Zona-Free Oocytes

Although Dickmann (1962) reported penetration of a single rat egg by a rabbit spermatozoon, it is widely accepted that spermatozoa of one species cannot fertilize eggs of another, except perhaps when the species are closely related and mating compatibility exists (e.g. sheep $\times$ goat). The zona pellucida acts as a formidable barrier to interspecific hybridization, primarily due to species specificity in its binding sites for spermatozoa (Chap. V), so removal of this acellular coat might permit the interaction of heterologous gametes: i.e. spermatozoa from one species might be able to fuse with the plasma membrane of eggs from another. At present, this seems to be generally not the case, for zona-free eggs from most mammals examined still retain relatively strong species specificity (Yanagimachi 1984). However, zona-free eggs of the golden hamster are peculiarly susceptible to interspecific gamete reactions and are able to incorporate spermatozoa of other mammalian species (Yanagimachi, Yanagimachi and Rogers 1976), even after cryopreservation and subsequent thawing (Fleming, Yanagimachi and Yanagimachi 1979). The underlying physiology is not fully understood, but capacitation is necessary for fusion to occur, presumably since the vesiculation reaction of the acrosome that follows from capacitation renders the equatorial and post-acrosomal regions of the sperm head attractive to the egg plasmalemma (see Chap. V). Sperm heads incorporated in this manner undergo nuclear decondensation and pronucleus formation in the egg cytoplasm, although only to a limited degree if there is excessive polyspermy.

In the light of this observation on sperm nuclear decondensation, one of the first proposals for the zona-free hamster oocyte was to use it for karyotyping the human spermatozoon — that is to examine the haploid chromosome complement — in situations in which genetic screening was deemed prudent (Yanagimachi et al. 1976; Rudak, Jacobs and Yanagimachi 1978; Yanagimachi 1982). Refinements of the technique used in these first studies to display the sperm chromosomes have been published (Martin 1983) and Chaudhuri and Yanagimachi (1984), and many of the results summarised and tabulated (Yanagimachi 1984).

Clinical applications of the zona-free hamster oocyte have been concerned with screening the male partner of infertile couples; there is evidence that the functional competence of human spermatozoa can be assessed to a limited extent by this method, as was previously suggested in the non-living, intact human oocyte test (Overstreet and Hembree 1976). As inferred above, spermatozoa to be screened need to be liberated from the seminal plasma by centrifugation and then capacitated and acrosome-reacted by preincubation in a suitable physiological medium. Freshly ejaculated human spermatozoa with intact acrosomes cannot attach to and penetrate zona-free hamster oocytes (Yanagimachi 1984). The proportion of oocytes penetrated by the prepared spermatozoa under defined conditions may give a useful measure of fertilizing ability, although this conclusion is not above controversy. Yanagimachi (1984) makes the point that because sperm penetration rates can be altered by changing technical procedures, negative results must be evaluated cautiously. He also notes that spermatozoa of some clinically infertile men can yield high rates of penetration. These limitations must remain a cause of concern. Again, whilst positive correlations exist between the ability to penetrate zona-free hamster oocytes on the one hand and intact human oocytes on the other (Overstreet et al. 1980), there are exceptions and they raise problems; it is, of course, individuals who are under consideration in the clinic.

As to application in domestic animals, the technique has been used in attempts to predict the fertility of males standing at artificial insemination centres. Despite initial optimism, there is now no persuasive evidence for a bull selection policy based on the zona-free oocyte test. Fertility as judged under large-scale insemination programmes does not correlate closely with measurements in the in vitro test. It is important to keep a sense of perspective in these in vitro studies in humans and farm animals and not to be seduced by the techniques themselves. The ability of spermatozoa to fertilize in vivo is, after all, not only an expression of capacitation and the acrosome reaction but also a measure of the sperm's ability to traverse the length of the genital tract and to survive the changing fluid environments encountered en route. The latter are not evaluated by the zona-free oocyte test.

Other Cellular and Nuclear Manipulations

There is not scope in this final section to be comprehensive; rather, it will attempt no more than brief comments on a series of contemporary research topics and point the way to appropriate literature. The extent to which many of the techniques that follow may have relevance to our own species is uncertain at the time of writing, and the cause of measured debate amongst informed parties. But it is not the intention of this volume to enter into controversy. Whether today's sensibilities will appear a trifle indulgent tomorrow remains to be seen.

In vitro maturation of oocytes is still the subject of intensive research in domestic animals (e.g. Moor and Crosby 1987), whereas the earlier work on human oocytes (e.g. Edwards 1965, 1966; Edwards, Bavister and Steptoe 1969) has been superseded by modern methods for estimating the maturity of follicles and pre-ovulatory oocytes. Because the ovaries contain their full complement of oocytes by the time of birth (Baker 1963, 1972), and since large numbers of primary oocytes can be liberated physically from Graafian follicles, resumption of meiosis in vitro was initially viewed as a means of dissociating oocyte maturation from the events of ovulation. It might also enable substantial numbers of secondary oocytes to be available for fertilization in vitro or in vivo, although the functional competence of many in vitro matured oocytes remains questionable (Leib-fried-Ruttledge et al. 1987). As of writing, however, studies of oocyte maturation seem to be more concerned with the nature of fundamental cellular changes, such as protein synthesis and membrane modifications, than with preparing oocytes for use in fertility studies. A notable finding in this work on oocyte maturation is the synthesis (or unmasking) of a cytoplasmic male pronucleus growth factor during the pre-ovulatory interval. This substance is not available in primary oocytes, and therefore decondensation of the sperm head into a pronuclear structure does not take place when immature eggs are exposed to spermatozoa in vitro (Thibault and Gérard 1970, 1973; Szollösi and Gérard 1983).

Cellular manipulations that have attracted attention in experimental animals include the physical separation of blastomeres at the 2- or 4-cell stage enabling the formation of identical (monozygotic) twins, triplets or even quadruplets; this requires the insertion of naked blastomeres into a 'recipient' zona pellucida. Earlier studies and their limitations were summarized by Hunter (1980), Willadsen and Polge (1980) and Polge (1985). Perhaps the most recent dramatic example of the power of these techniques is the production of five identical (quintuplet) lambs from five blastomeres disaggregated from an 8-cell embryo (Willadsen and Fehilly 1983). Such a demonstration raises major questions concerning the stage to which totipotency is retained (or can be regained) by the nuclei of individual cells in the cleaving embryo. Irreversible differentiation had clearly not occurred in a majority of the blastomeres by the 8-cell stage. Techniques for producing monozygotic quadruplets in cattle have also been described (Willadsen and Polge 1981).

Turning to chimaeras — composite animals in which different cell populations are derived from more than one zygote either by aggregating denuded embryos or by injecting embryonic cells into blastocysts — these were first produced experimentally in mice (Tarkowski 1961; Mintz 1962; Gardner 1968; McLaren 1976). The technique of chimaera formation is an elegant way of studying problems of differentiation and de-differentiation by examining the fate of the composite cell lines at different stages of development; it might also yield valuable information on malignancy and on possible ways of controlling the growth of tumours. As far as applications in farm animals are concerned, formation of chimaeras has been proposed as a means of incorporating desirable traits from one breed into another, in reality producing new strains or breeds of animal. Perhaps the most widely noted example to date has been the formation of a sheep-goat chimaera (Fehilly, Willadsen and Tucker 1984). Comments as to any possible clinical relevance of these techniques in human medicine would be out of place in this volume.

This discussion so far has focussed on manipulations at the cellular level. Manipulations at the nuclear level include limited success in the formation of clones by means of nuclear transplantation (reviewed by Markert 1984; McLaren 1982, 1984; Gurdon 1986), but the earlier work of Gurdon (1968) in amphibia was more fruitful than the studies so far reported in mammals. He introduced the nucleus of a differentiated somatic cell into the cytoplasm of an unfertilized egg whose own nucleus had been removed or destroyed, and was thereby able to obtain normal development. Whilst cloning could clearly be of value in farm animals (e.g. clones of high-yielding dairy cows), of much greater clinical relevance in the future may be the technique of microinjection of single sperm heads (i.e. essentially the sperm nucleus) directly into the egg cytoplasm (Markert 1983). Whether this is viewed as a cellular rather than nuclear manipulation is open to debate. Such microinjection might offer a means of overcoming infertility in our own species due to severe oligospermy or extremely poor sperm motility. But much remains to be learnt about such techniques, not least the required state of the sperm head membranes at the time of injection that would enable egg activation and nuclear decondensation. More specifically, does the acrosome reaction need to be induced prior to injection in order for subsequent formation of a male pronucleus?

Other uses of nuclear transplantation have been to examine compatibility of nuclear stage with that of the cytoplasm, by fusing nuclei taken from cells of different developmental ages with enucleated blastomeres or even oocytes; the fusion is mediated by inactivated Sendai virus or dielectrophoresis. Studies along these lines have been successful in mice (e.g. Surani, Barton and Norris 1987), as had earlier studies with nuclear injection that were not combined with prior enucleation of the recipient zygotes (Modlinski 1978). More remarkable were the experimental findings of Hoppe and Illmensee (1977, 1982) in which removal of one of the pronuclei and diploidization of the remaining pronucleus was reported to give a viable, homozygous, uniparental mouse. The inability of other groups to repeat these findings has caused concern (see Markert 1984). Nuclear transplantation between sheep embryos has been achieved using the blastomere of an 8-cell embryo to effect nuclear combination with the cytoplasm of an unfertilized egg. Such reconstituted embryos can develop into viable lambs (Willadsen 1986).

Whilst nuclear transplantation may enable control of the total genome, modification of the genetic status of an embryo by DNA injection has attracted the label of genetic engineering and can result in transgenic animals (Hammer et al. 1985). The technique involves microinjection of appropriate DNA fragments or constructs into one of the pronuclei during the early stages of fertilization; the male pronucleus is invariably chosen since it is larger than the female. DNA replication occurs within the migrating pronuclei (Oprescu and Thibault 1965; Szöllösi 1966) since, when the chromosomes condense at syngamy, they enter straight into the first mitotic division. DNA injected into a pronucleus may therefore be integrated into the genome and subsequently expressed. This was dramatically demonstrated to be the case when genes for growth hormone were injected into pronuclear mouse eggs, enabling generation of giant animals (Palmiter et al. 1982). The prospects for farm animals have also been reviewed (Wagner et al. 1984; Simons and Land 1987). Whether gene therapy proves to be a practical and reliable means of overcoming genetical errors in our own species is controversial, although there is little doubt that studies in experimental animals will continue to explore DNA therapy and, in due course, manipulation of specific individual genes.

This chapter would not be complete without reference to methods of predetermination or selection of sex. Y-bearing spermatozoa may be identified by means of an appropriate DNA probe, giving the potential for selection of either X- or Y-bearing gametes; these might then be used through procedures of artificial insemination or in vitro fertilization. As to pre-implantation embryos, males can be revealed by the use of a Y-probe and females can be recognized since one of the X-chromosomes of somatic cells becomes inactivated, giving rise to a lump of chromatin − the sex chromatin or Barr body. This can be detected in the nucleus upon staining. Embryos can therefore be screened by removing a small portion of trophoblast, a manipulation that does not compromise the viability of the embryo. This technique was first demonstrated in laboratory animals by Gardner and Edwards (1968), enabling successful prediction of sex in blastocysts subsequently transferred to recipient animals.

Finally, a word about freeze-preservation of cells, the science of cryobiology, which is very much a research field in its own right. Cryopreservation of gametes and/or embryos is a technology that has already conferred immense benefits in the breeding of livestock, both in terms of the artificial insemination industry and for the transplantation of embryos. The observation that glycerol could be used to protect sperm cells from the deleterious effects of freezing and thawing (Polge, Smith and Parkes 1949) led to detailed investigations by biophysicists also interested in the freeze-preservation of red blood cells. Intra-cellular formation of ice was identified as a major problem, causing disruption of the cell organelles and cytoskeleton during thawing, but cryoprotective agents such as glycerol or dimethyl sulphoxide have since been used with great success. Many of the requirements of a system of freezing and thawing have been accurately defined, such as the concentration of protective agent in the medium, the time and temperature of equilibration, the nature of the cooling and thawing curves, and the means of removal of the protective agent. This technical background, based on studies of animal embryos (see Whittingham 1976; Leibo and Mazur 1978; Polge and Willadsen 1978) clearly facilitated the successful freeze-preservation and subsequent transfer of human embryos (e.g. Mohr and Trounson 1983; Zeilmaker et al. 1984). The latter steps have been practised upon what are termed spare embryos arising from multiple ovulations in women treated at fertility clinics. Once again, there are considerable misgivings concerning the application of cryopreservation techniques to human embryos. The arguments have been well rehearsed and do not require repetition here.

In the field of animal conservation, deep-frozen storage of gametes and embryos represents an important method of genome conservation, and one which may be regarded as an insurance policy permitting the preservation and future reconstitution of unique genotypes (Polge 1985). Such reconstitution would necessarily require transplantation into the Fallopian tubes or uterus of suitable recipients as the final step leading to a fruitful outcome.

REFERENCES

Allen NC, Herbert CM, Maxson WS, Rogers BJ, Diamond MP, Wentz AC (1985) Intrauterine insemination: a critical review. Fertil Steril 44:569–580
Asch RH, Ellsworth LR, Balmaceda JP, Wong PC (1984) Pregnancy after translaparoscopic gamete intrafallopian transfer. Lancet ii:1034–1035
Asch RH, Ellsworth LR, Balmaceda JP, Wong PC (1985) Birth following gamete intrafallopian transfer. Lancet ii:163
Austin CR (1951) Observations on the penetration of the sperm into the mammalian egg. Aust J Sci Res (B) 4:581–596
Austin CR (1961a) The mammalian egg. Blackwell Sci Publ, Oxford
Austin CR (1961b) Fertilization of mammalian eggs in vitro. Int Rev Cytol 12:337–359
Baker RD, Dziuk PJ (1970) Aerial transport of fertilized pig ova. Canad J Anim Sci 50:215–216
Baker TG (1963) A quantitative and cytological study of germ cells in human ovaries. Proc R Soc London, Ser B 158:417–433
Baker TG (1972) Primordial germ cells. In: Austin CR, Short RV (eds) Reproduction in mammals, book 1. Cambridge University Press, pp 1–13
Ball GD, Leibfried ML, Lenz RW, Ax RL, Bavister BD, First NL (1983) Factors affecting successful in vitro fertilisation of bovine follicular oocytes. Biol Reprod 28:717–725
Barros C, Austin CR (1967) In vitro fertilization and the sperm acrosome reaction in the hamster. J Exp Zool 166:317–324
Bavister BD (1969) Environmental factors important for in vitro fertilization in the hamster. J Reprod Fertil 18:544–545
Beier HM, Lindner HR (eds) (1983) Fertilization of the human egg in vitro. Springer, Berlin Heidelberg New York
Betteridge KJ (1977) Embryo transfer in farm animals. Monograph No 16, Can Dept Agric, Ottawa
Betteridge KJ (1986) Increasing productivity in farm animals. In: Austin CR, Short RV (eds) Reproduction in mammals, vol 5. Cambridge University Press, pp 1–47
Biggers JD (1981) In vitro fertilisation and embryo transfer in human beings. New Engl J Med 304:336–342

Bomsel-Helmreich O (1985) Ultrasound and the preovulatory human follicle. In: Clark JR (ed) Oxford reviews of reproductive biology, vol 7. Clarendon Press, Oxford, pp 1—72

Brackett BG (1981) Applications of in vitro fertilization. In: Brackett BG, Seidel GE, Seidel SM (eds) New technologies in animal breeding. Academic Press, New York, pp 141—161

Brackett BG, Bousquet D, Boice ML, Donawick WJ, Evans JF, Dressel MA (1982) Normal development following in vitro fertilisation in the cow. Biol Reprod 27:147—158

Chang MC (1949) Effects of heterologous sera on fertilized rabbit ova. J Gen Physiol 32:291—300

Chang MC (1950) Development and fate of transferred rabbit ova or blastocyst in relation to the ovulation time of recipients. J Exp Zool 114:197—225

Chang MC (1951) Fertilizing capacity of spermatozoa deposited into the Fallopian tubes. Nature (Lond) 168:697—698

Chang MC (1959) Fertilization of rabbit ova in vitro. Nature (Lond) 184:466—467

Chang MC (1968) In vitro fertilization of mammalian eggs. J Anim Sci 27 Suppl 1:15—22

Chaudhuri JP, Yanagimachi R (1984) An improved method to visualize human sperm chromosomes using zona-free hamster eggs. Gamete Res 10:233—239

Cheng WT (1985) Doctoral dissertation. CNAA

Daniel JC (1978) Methods in mammalian reproduction. Academic Press, New York

Dauzier L, Thibault C (1956) Recherche expérimentale sur la maturation des gamètes males chez les mammifères par l'étude de la fécondation 'in vitro' de l'oeuf de lapine. Proc 3rd Int Congr Anim Reprod, Cambridge 1:58—61

Dauzier L, Thibault C, Wintenberger S (1954) La fécondation in vitro de l'oeuf de la lapine. CR Acad Sci, Paris 238:844—845

Dellenbach P, Nisand I, Moreau L, Feger B, Plumere C, Gerlinger P (1985) Transvaginal sonographically controlled follicle puncture for oocyte retrieval. Fertil Steril 44:656—662

Demoulin A (1985) Is diagnostic ultrasound safe during the periovulatory period? Res Reprod 17:1—2

Dickmann Z (1962) Experiments on interspecific sperm penetration through the zona pellucida. J Reprod Fertil 4:121—124

Edwards RG (1965) Maturation in vitro of mouse, sheep, cow, rhesus monkey and human ovarian oocytes. Nature (Lond) 208:349—351

Edwards RG (1966) Mammalian eggs in the laboratory. Sci Am 215:73—81

Edwards RG (1980) Conception in the human female. Academic Press, London

Edwards RG (1984) The current situation of in vitro fertilization. IPPF Med Bull 18:1—2

Edwards RG, Donahue RP, Baramki TA, Jones HW (1966) Preliminary attempts to fertilize human oocytes matured in vitro. Am J Obstet Gynecol 96:192—200

Edwards RG, Bavister BD, Steptoe PC (1969) Early stages of fertilisation in vitro of human oocytes matured in vitro. Nature (Lond) 221:632—635

Edwards RG, Steptoe PC, Purdy JM (1970) Fertilization and cleavage in vitro of preovulatory human oocytes. Nature (Lond) 227:1307—1309

Edwards RG, Steptoe PC, Purdy JM (1980) Establishing full-term human pregnancies using cleaving embryos grown in vitro. Brit J Obstet Gynaecol 87:737—756

Edwards RG, Steptoe PC, Fowler RE, Baillie J (1987) Observations on preovulatory human ovarian follicles and their aspirates. Brit J Obstet Gynaecol 87:769—779

Fehilly CB, Willadsen SM, Tucker EM (1984) Interspecific chimaerism between sheep and goat. Nature (Lond) 307:634—636

Feichtinger W, Kemeter P (eds) (1987) Future aspects in human in vitro fertilization. Springer, Berlin Heidelberg New York Tokyo

Fergusson ILC (1982) Laparoscopic investigation of tubal infertility. In: Chamberlain G, Winston R (eds) Tubal infertility: diagnosis and treatment. Blackwell, Oxford, pp 30—46

Fleming AD, Yanagimachi R, Yanagimachi H (1979) Fertilisability of cryopreserved zona-free hamster ova. Gamete Res 2:357—366

Forrler A, Dellenbach P, Nisand I, Moreau L, Cranz C, Clavert A, Rumpler Y (1986a) Direct intraperitoneal insemination in unexplained and cervical infertility. Lancet i:916—917

Forrler A, Badoc E, Moreau L, Dellenbach P, Cranz C, Clavert A, Rumpler Y (1986b) Direct intraperitoneal insemination: first results confirmed. Lancet ii:1468

Frydman R, Testart J, Giacomini P, Imbert MC, Martin E, Nahoul K (1982) Hormonal and histological study of the luteal phase in women following aspiration of the preovulatory follicle. Fertil Steril 38:312–317

Garcia J, Jones GS, Acosta AA, Wright GL (1981) Corpus luteum function after follicle aspiration for oocyte retrieval. Fertil Steril 36:565–572

Gardner RL (1968) Mouse chimaeras obtained by the injection of cells into the blastocyst. Nature (Lond) 220:596–597

Gardner RL, Edwards RG (1968) Control of the sex ratio at full term in the rabbit by transferring sexed blastocysts. Nature (Lond) 218:346–348

Gurdon JB (1968) Transplanted nuclei and cell differentiation. Sci Am 219:24–35

Gurdon JB (1986) Nuclear transplantation in eggs and oocytes. J Cell Sci Suppl 4:287–318

Gwatkin RBL (1977) Fertilization mechanisms in man and mammals. Plenum Press, New York

Hadek R (1958) Intraperitoneal insemination of rabbit doe. Proc Soc Exp Biol Med 99:39–40

Hammer RE, Pursel VG, Rexroad CE, Wall RJ, Bolt DJ, Ebert KM, Palmiter RD, Brinster RL (1985) Production of transgenic rabbits, sheep and pigs by microinjection. Nature (Lond) 315:680–683

Harms von E, Smidt D (1970) In vitro fertilization of follicular and tubal eggs of swine. Tierärztl Wochenschr 83:269–275

Harper MJK, Bennet JP, Rowson LEA (1961) A possible explanation for the failure of non-surgical ovum transfers in the cow. Nature (Lond) 190:789–790

Hartmann JF (1983) Mammalian fertilization: gamete surface interactions in vitro. In: Hartmann JF (ed) Mechanism and control of animal fertilization. Academic Press, New York, pp 325–364

Hoppe PC, Illmensee K (1977) Microsurgically produced homozygous-diploid uniparental mice. Proc Natn Acad Sci USA 74:5657–5661

Hoppe PC, Illmensee K (1982) Full-term development after transplantation of parthenogenetic embryonic nuclei into fertilized mouse eggs. Proc Natn Acad Sci USA 79:1912–1916

Hunter RHF (1967) Polyspermic fertilization in pigs during the luteal phase of the estrous cycle. J Exp Zool 165:451–460

Hunter RHF (1978) Intraperitoneal insemination, sperm transport and capacitation in the pig. Anim Reprod Sci 1:167–179

Hunter RHF (1980) Physiology and technology of reproduction in female domestic animals. Academic Press, London

Hunter RHF (1987) The timing of capacitation in mammalian spermatozoa – a reinterpretation. Res Reprod 19:3–4

Hunter RHF, Lawson RAS, Rowson LEA (1972) Maturation, transplantation and fertilization of ovarian oocytes in cattle. J Reprod Fertil 30:325–328

Iritani A, Niwa K, Imai H (1978) Sperm penetration in vitro of pig follicular oocytes matured in culture. J Reprod Fertil 54:379–383

Iritani A, Kasai M, Niwa K, Song HB (1984) Fertilisation in vitro of cattle follicular oocytes with ejaculated spermatozoa capacitated in a chemically defined medium. J Reprod Fertil 70:487–492

Irvine DS, Aitken RJ, Lees MM, Reid C (1986) Failure of high intrauterine insemination of husband's semen. Lancet ii:972–973

Jones HW, Jones GS, Hodgen GD, Rosenwaks Z (eds) (1986) In vitro fertilization. Williams Wilkins, Baltimore

Kerin JF, Broom TJ, Ralph MM, Edmonds DK, Warner GM, Jeffrey R, Crocker JM, Godfrey B, Cox LW, Seamark RF, Mathews CD (1981) Human luteal phase function following oocyte aspiration from the immediate preovular Graafian follicle of spontaneous ovulating cycles. Brit J Obstet Gynaecol 88:1021–1028

Kerin JFP, Peek J, Warnes GM, Kirby C, Jeffrey R, Mathews CD, Cox LW (1984) Improved conception rate after intrauterine insemination of washed spermatozoa from men with poor quality semen. Lancet i:533–535

Land RB, Morris BA, Baxter G, Fordyce M, Forster J (1982) Improvement of sheep fecundity by treatment with antisera to gonadal steroids. J Reprod Fertil 66:625–634

Leibfried-Rutledge ML, Critser ES, Eyestone WH, Northey DL, First NL (1987) Development potential of bovine·oocytes matured in vitro or in vivo. Biol Reprod 36:376–383

Leibo SP, Mazur P (1978) Methods for the preservation of mammalian embryos by freezing. In: Daniel JC (ed) Methods in mammalian reproduction. Academic Press, New York, pp 179 – 201

Mann T (1964) The biochemistry of semen and of the male reproductive tract. Methuen, London

Markert CL (1983) Fertilization of mammalian eggs by sperm injection. J Exp Zool 228:195 – 201

Markert CL (1984) Cloning mammals: current reality and future prospects. Theriogenology 21:60 – 67

Martin RH (1983) A detailed method for obtaining preparations of human sperm chromosomes. Cytogenet Cell Genet 35:252 – 256

McLaren A (1976) Mammalian chimaeras. Cambridge University Press, London

McLaren A (1982) The embryo. In: Austin CR, Short RV (eds) Reproduction in mammals, 2nd eds. Cambridge University Press, pp 1 – 25

McLaren A (1984) Methods and success of nuclear transplantation in mammals. Nature (Lond) 309:671 – 672

McNeilly AS (1985) Effect of changes in FSH induced by bovine follicular fluid and FSH infusion in the preovulatory phase on subsequent ovulation rate and corpus luteum function in the ewe. J Reprod Fertil 74:661 – 668

Messinis IE, Templeton AA (1987) Disparate effects of endogenous and exogenous oestradiol on luteal phase function in women. J Reprod Fertil 79:549 – 554

Mintz B (1962) Formation of genotypically mosaic mouse embryos. Am Zool 2:432

Modlinski JA (1978) Transfer of embryonic nuclei to fertilised mouse eggs and development of tetraploid blastocysts. Nature (Lond) 273:466 – 467

Mohr L, Trounson A (1983) Human pregnancy following cryopreservation, thawing and transfer of an eight-cell embryo. Nature (Lond) 305:707 – 709

Molloy D, Speirs A, du Plessis Y, Gellert S, Bourne H, Johnston WIH (1986) The establishment of a successful programme of 'GIFT'. Austral NZ J Obstet Gynaecol 26:206 – 209

Moor RM, Crosby IM (1987) Cellular origin, hormonal regulation and biochemical characteristics of polypeptides secreted by Graafian follicles of sheep. J Reprod Fertil 79:469 – 483

Oprescu ST, Thibault C (1965) Duplication de l'ADN dans les oeufs de lapine après la fécondation. Annls Bio Anim Biochim Biophys 5:151 – 156

Overstreet JW, Hembree WC (1976) Penetration of the zona pellucida of non-living human oocytes by human spermatozoa in vitro. Fertil Steril 27:815 – 831

Overstreet JW, Yanagimachi R, Katz DF, Hayashi K, Hanson FW (1980) Penetration of human spermatozoa into human zona pellucida and the zona-free hamster egg. A study of fertile donors and infertile patients. Fertil Steril 33:534 – 542

Palmiter RD, Brinster RL, Hammer RE, Trumbauer ME, Rosenfeld MG, Birnberg NC, Evans RM (1982) Dramatic growth of mice that develop from eggs microinjected with metallothionein-growth hormone fusion genes. Nature (Lond) 300:611 – 615

Polge C (1972) Increasing reproductive potential in farm animals. In: Austin CR, Short RV (eds) Reproduction in mammals, book 5. Cambridge University Press, Ch 1, pp 1 – 31

Polge C (1985) Embryo manipulation and genetic engineering. Symp Zool Soc Lond No 54:123 – 135

Polge C, Willadsen SM (1978) Freezing eggs and embryos of farm animals. Cryobiology 15:370 – 373

Polge C, Smith AU, Parkes AS (1949) Revival of spermatozoa after vitrification and dehydration at low temperatures. Nature (Lond) 164:666

Rolland R, Heineman MJ, Hillier SG, Vemer H (eds) (1985) Gamete quality and fertility regulation. Excerpta Medica, Amsterdam

Rowlands IW (1957) Insemination of the guinea pig by intraperitoneal injection. J Endocrinol 16:98 – 106

Rowlands IW (1958) Insemination by intraperitoneal injection. Proc Soc Stud Fertil 10: 150 – 157

Rowson LEA, Bennet JP, Harper MJK (1964) The problem of non-surgical egg transfer to the cow uterus. Vet Rec 76:21 – 23

Rudak E, Jacobs PA, Yanagimachi R (1978) Direct analysis of the chromosome constitution of human spermatozoa. Nature (Lond) 274:911−912

Schenk SL (1878) Das Säugetierei künstlich befruchtet außerhalb des Muttertieres. Mitt Embryol Inst Univ Wien 1:107−118

Seitz HM, Brackett BG, Mastroianni L (1970) In vitro fertilization of ovulated rabbit ova recovered from the ovary. Biol Reprod 2:262−267

Serhal PF, Katz M (1987) Intrauterine insemination. Lancet i:52−53

Simons JP, Land RB (1987) Transgenic livestock. J Reprod Fertil Suppl 34:237−250

Skjerven O (1955) Conception in a heifer after deposition of semen in the abdominal cavity. Fertil Steril 6:66−67

Staigmiller RB, Moor RM (1984) Effect of follicle cells on the maturation and developmental competence of ovine oocyes matured outside the follicle. Gamete Res 9:221−229

Steptoe PC (1977) Laparoscopy in gynaecology. In: Stallworthy JS, Bourne G (eds) Recent advances in obstetrics and gynaecology, No 12. Livingstone, Edinburgh

Steptoe PC, Edwards RG (1976) Reimplantation of a human embryo with subsequent tubal pregnancy. Lancet i:880−882

Steptoe PC, Edwards RG (1978) Birth after the reimplantation of a human embryo. Lancet ii:366

Steptoe PC, Webster J (1982) Laparoscopy of the normal and disordered ovary. In: Edwards RG, Purdy JM (eds) Human conception in vitro. Academic Press, New York

Steptoe PC, Edwards RG, Walters DE (1986) Observations on 767 clinical pregnancies and 500 births after human in-vitro fertilization. Human Reprod 1:89−94

Surani MAH, Barton SC, Norris ML (1987) Experimental reconstruction of mouse eggs and embryos: an analysis of mammalian development. Biol Reprod 36:1−16

Szöllösi D (1966) Time and duration of DNA synthesis in rabbit eggs after sperm penetration. Anat Rec 154:209−212

Szöllösi D, Gérard M (1983) Cytoplasmic changes in the mammalian oocytes during the preovulatory period. In: Beier HM, Lindner HR (eds) Fertilization of the human egg in vitro. Springer, Berlin Heidelberg New York, pp 35−55

Tarkowski AK (1961) Mouse chimaeras developed from fused eggs. Nature (Lond) 190:857−860

Testart J, Frydman R (eds) (1985) Human in vitro fertilization. Elsevier, Amsterdam

Thibault C (1969) In vitro fertilization of the mammalian egg. In: Metz CB, Monroy A (eds) Fertilization. Comparative morphology, biochemistry and immunology, vol 2. Academic Press, London, pp 405−435

Thibault C, Dauzier L (1960) 'Fertilisines' et fécondation in vitro de l'oeuf de lapine. CR Acad Sci, Paris 250:1358−1359

Thibault C, Dauzier L (1961) Analyse des conditions de la fécondation in vitro de l'oeuf de la lapine. Annls Biol Anim Biochim Biophys 1:277−294

Thibault C, Gérard M (1970) Facteur cytoplasmique nécessaire à la formation du pronucleus mâle dans l'ovocyte de lapine. CR Acad Sci, Paris 270:2025−2026

Thibault C, Gérard M (1973) Cytoplasmic and nuclear maturation of rabbit oocytes in vitro. Annls Biol Anim Biochim Biophys 13:145−156

Thomas EJ, McTighe L, King H, Lenton EA, Harper R, Cooke ID (1986) Failure of high intrauterine insemination of husband's semen. Lancet ii:693−694

Toffle RC, Nagel TC, Tagatz CE, Phansey SA, Okagaki T, Wavrin CA (1985) Intrauterine insemination: the University of Minnesota experience. Fertil Steril 43:743−747

Trounson AO, Mohr LR, Wood C, Leeton JF (1982) Effect of delayed insemination on in vitro fertilization, culture and transfer of human embryos. J Reprod Fertil 64:285−294

Trounson A, Hoppen HO, Lutjen PJ, Mohr LR, Rogers PAW, Sathananthan AH (1985) In vitro fertilization: the state of the art. In: Rolland R, Heineman MJ, Hillier SG, Vemen H (eds) Gamete quality and fertility regulation. Excerpta Medica, Amsterdam, pp 325−343

Wagner TE, Murray F, Minhas B, Kraemer DC (1984) The possibility of transgenic livestock. Theriogenology 21:29−44

Wallace J, McNeilly A (1986) Changes in FSH and the pulsatile secretion of LH during treatment of ewes with bovine follicular fluid throughout the luteal phase of the oestrous cycle. J Endocrinol 111:317−327

Webb R (1987) Increasing ovulation rate and lambing rate in sheep by treatment with a steroid enzyme inhibitor. J Reprod Fertil 79:231–240

Whittingham DG (1976) General aspects of egg culture and preservation. In: Rowson LEA (ed) Egg transfer in cattle. Proc EEC Res Seminar, pp 101–113

Wikland M, Nilsson L, Hansson R, Hamberger L, Janson PO (1983) Collection of human oocytes by the use of sonography. Fertil Steril 39:603–608

Willadsen SM (1986) Nuclear transplantation in sheep embryos. Nature (Lond) 320:63–65

Willadsen SM, Fehilly CB (1983) The developmental potential and regulatory capacity of blastomeres from two-, four- and eight-cell sheep embryos. In: Beier HM, Lindner HR (eds) Fertilization of the human egg in vitro. Springer, Berlin Heidelberg New York, pp 353–357

Willadsen SM, Polge C (1980) Embryo transplantation in the large domestic species. J Roy Agric Soc 141:115–126

Willadsen SM, Polge C (1981) Attempts to produce monozygotic quadruplets in cattle by blastomere separation. Vet Rec 108:211–213

Wilmut I, Sales DI, Ashworth CJ (1986) Maternal and embryonic factors associated with prenatal loss in mammals. J Reprod Fertil 76:851–864

Wolf DP, Byrd W, Dandekar P, Quigley M (1984) Sperm concentration and the fertilization of human eggs in vitro. Biol Reprod 31:837–848

Wood C, Trounson A (eds) (1984) Clinical in vitro fertilization. Springer, Berlin Heidelberg New York Tokyo

Wright RW, Bondioli KR (1981) Aspects in vitro fertilization and embryo culture in domestic animals. J Anim Sci 53:702–729

Yanagimachi R (1982) Potential methods for examining sperm chromosomes. In: Amann RP, Seidel GE (eds) Prospects for sexing mammalian sperm. Colorado University Press, Boulder, pp 225–251

Yanagimachi R (1984) Zona-free hamster eggs: their use in assessing fertilizing capacity and examining chromosomes of human spermatozoa. Gamete Res 10:187–232

Yanagimachi R, Chang MC (1963) Fertilization of hamster eggs in vitro. Nature (Lond) 200:281–282

Yanagimachi R, Chang MC (1964) In vitro fertilization of golden hamster ova. J Exp Zool 156:361–376

Yanagimachi R, Yanagimachi H, Rogers BJ (1976) The use of zona-free animal ova as a test system for the assessment of the fertilizing capacity of human spermatozoa. Biol Reprod 15:471–476

Yovich JL, Matson PL (1986) Pregnancy rates after high intrauterine insemination of husband's spermatozoa or gamete intrafallopian transfer. Lancet ii:1287

Zeilmaker GH, Alberda AT, van Gent I, Rikmans CP, Drogendijk AC (1984) Two pregnancies following transfer of intact frozen-thawed embryos. Fertil Steril 42:293–296

CHAPTER X

Are the Fallopian Tubes Essential?
A Biological Perspective

These final pages are intended to be brief — in fact scarcely a chapter, just a few paragraphs. Their purpose concerns a meaningful interpretation of the overall function of the Fallopian tubes and, more specifically, the rôle of individual components of the tubal environment. Throughout the preceding text, the question has repeatedly arisen as to whether features that appear peculiar to the Fallopian tubes — or are highlighted therein — are essential to mammalian reproduction. Based upon experimental evidence from individual animals, the answer has all too often been an unequivocal no. For example, the contribution of the myosalpinx to the process of egg transport has been questioned in situations in which perturbed contractions have not prevented fertility. Similarly, a specific involvement of the cilia in egg transport has been questioned in instances where pregnancies have been established in the absence of such structures. The list of such potential divergences would be repetitive of the text if cited in full. However, the most dramatic example of usurping the normal physiological functions of the Fallopian tube remains that of in vitro fertilization of human ovarian oocytes and subsequent transplantation of the embryo directly into the uterus; pregnancies can be established without any contact between gametes and the tubal structures.

In searching for a biological perspective, to accept the above findings at face value is perhaps to miss the point. For most biologists, successful reproduction needs to be considered in population terms, and all features that contribute to the stability of a breeding population, and thereby to perpetuation of the species, need to be brought to mind. It is in this light that the many specific contributions of the Fallopian tubes are best interpreted. Whilst individual morphological and biochemical components may not be of overriding importance in themselves for the establishment of single pregnancies, together they can be viewed as contributing to the successful maintenance of a breeding population.

Analogous arguments can be mounted for the male. Classical studies in the 1930s revealed that the accessory glands of the reproductive tract — the prostate, seminal vesicles and Cowper's glands — could be extirpated gland by gland and yet the surgically-modified male could still generate pregnancies, albeit at a much lower incidence than in the intact condition. The results from artificial insemination of washed ejaculated spermatozoa or suspensions of epididymal spermatozoa make a similar point, although insemination deeper into the female tract than occurs at mating tends to compensate and enable high conception rates to be obtained. Few would argue from this evidence that the secretions of the male accessory glands had only a vestigial rôle in reproductive processes. Rather, such secretions would seem of value for promoting conception in the range of females expected to constitute a breeding population.

These well-rehearsed arguments may be reasonable and appropriate as a conventional conclusion, enabling us to view the various specializations of the Fallopian tubes with some sensitivity. Nonetheless, the suspicion lingers that interpretation of reproductive processes in primates and, in particular, tubal function in women, may require an altogether different plane of thought. As reasoned in diverse texts for diverse audiences down the ages, the human embryo may indeed be unique both in features of its generation and expression of its constitution.

Subject Index

abdominal surgery 9
– window 8, 100
abortion 118, 143, 171
acrosin 93
acrosomal enzymes 72, 83, 89, 92, 95, 99
– membranes 70, 89, 93, 97
acrosome 83, 84, 89, 93, 173
– reaction 84, 85, 88, 89, 92–94, 97, 173–175
actin 99
activation 53, 101
adhesions 140, 144
adrenergic blockers 135
– fibres 22, 23, 54, 131
– nerve terminals 22, 54, 131
– receptors 131
ageing 58
albumin 37, 92, 113, 164, 172
amino acids 36, 38, 39, 113, 115
ampulla 12, 13, 16–24, 26, 33, 34, 37, 41, 43–47, 54, 57, 61, 63, 67–72, 92, 128, 136, 140, 145, 153, 164, 171
ampullary-isthmic junction 12, 13, 22, 26, 44, 53, 57, 66, 67, 82, 128–130, 151, 152
androgens 14, 140
androstenedione 46
aneuploids 101, 102
antibodies 72
antifertilizin 84, 85
antigens 111
anti-Müllerian hormone 14
antral fluid 57
arterioles 18, 19, 21
artificial eggs 8
– insemination 53, 60, 68, 70, 172, 176
ATP 95
autonomic nervous system 22, 23, 54, 57, 68, 81, 135

baboon 37, 130
Baer, von 7
bat 103, 110, 111, 130

bicarbonate 39, 41, 43, 44, 83, 113, 163, 169
bicornuate uterus 2
biochemical composition 8
– techniques 7
bladder 2
blastocoele 121
blastocysts 4, 110, 112, 115, 116, 119–122, 145, 147, 169, 175
blastomeres 111, 112, 120, 121, 174–176
block to polyspermy 97, 99, 101, 102, 112
blood flow 18, 30, 45
breeding success 119
broad ligament 20

calcium ions 92, 95, 99, 164
– transport 48, 88, 92
calmodulin 92
camels 58
capacitation 39, 61, 70, 71, 88–93, 95, 147–149, 161–164, 168, 172–174
capillary bed 18, 19, 115
carnivore 110
cat 26, 57, 67
catecholamines 81, 91, 131
catheterisation 33, 36
cell density 59, 60, 64
– junctions 121
– lines 120, 121
– surface 111
– types 111
cervical canal 59, 60
– crypts 59, 60
– mucus 59, 60, 172
– os 58, 59
cervix 15, 58–63, 91, 148, 150, 170
chemotaxis 70, 81, 84
chiasma frequency 71
chimaeras 175
cholinergic innervation 23
chorionic gonadotrophin 116
chromatin 95, 97, 98
– decondensation 97

chromosome anomalies   14, 97, 101, 102
chromosomes   97, 99, 101, 121, 167, 173, 176
cilia   8, 14, 24, 26, 48, 54–58, 67, 70, 83, 130–132, 140, 143, 151, 171, 183
cilial tips   58
ciliary activity   7, 26, 48, 55, 57, 58, 67, 70, 140, 171
– beat   26, 48, 57, 67, 140, 144, 151, 171
– currents   34
ciliated epithelium   55, 57, 58, 70
circular muscle   66, 131, 135
cleavage   30, 39, 48, 99–101, 109, 111–115, 120–122, 168, 169
clomiphene citrate   165, 166
cloning   175
CO_2 43, 113, 169
coitus   61, 112
conception rate   30
conceptus   112
congenital disturbances   140
connective tissue   19
contraceptive therapy   30, 128, 135, 143
contractile activity   7, 48, 57, 61, 67, 69, 83, 130, 131, 134
contractions of tube   7, 20, 48, 57, 131
copulation   4, 59
corona cells   39, 82, 83, 86
– radiata   54–56, 82, 83
corpora lutea   18, 73, 116, 119, 127, 131, 135, 144, 145, 165, 167
cortical granules   74, 97–99, 101
counter-current transfer   19, 20, 47, 69
cow   8, 12, 13, 18, 24, 26, 31–33, 36, 37, 41, 54, 56, 58, 61, 62, 67, 70, 82, 84, 90, 92, 100, 110, 117, 120, 129, 132, 162, 163, 169, 171, 175
Cruickshank, William   4
cryobiology   33, 176
cryopreservation   173, 177
culture in vivo   120
cumulus cells   56, 72, 82–84, 93, 113, 147, 148, 163, 164, 167
– mass   7, 8, 26, 54, 56, 57, 72, 82–84, 100, 151, 163, 167
cytoplasmic degeneration   103
– droplet   72
– processes   54, 55
cytoskeleton   97, 99, 176

decapacitation   91
deciliation   24, 26
deer   4
defective spermatozoa   71
deutoplasm   112
differentiation   111, 121, 122, 175
digyny   101

dimensions   12, 13
diploid spermatozoa   71, 101
disulphide bridges   95
DNA   99, 122, 176
dog   58, 62, 63, 68, 73, 89
dynein arms   58

early pregnancy factors   8, 19, 20, 45, 48, 110, 115–119
echo-sonography   83, 110, 167
ectopic pregnancy   54, 58, 135, 136, 141–145, 152, 153, 156, 170, 171
egg   4, 5, 7, 9, 24, 26, 37, 53–58, 60, 62, 67, 68, 71–74, 81, 82, 85, 90, 91, 99, 100, 109, 128, 129, 148, 161, 162
– activation   97, 100, 101, 113, 161
– ageing   53, 58, 72, 74, 101–103
– denudation   83
– investments   58, 67, 70, 71, 81, 90, 93, 95, 102
– membranes   57, 59, 70, 72, 81, 88, 96
– retention   131, 143
– surface   113, 132
– transplantation   67, 111, 113, 127, 147, 148, 161, 162, 168
– transport   7, 8, 26, 53, 54, 57, 68, 85, 110, 127–135, 143, 144, 183
ejaculate   60, 72, 89, 90, 102, 150, 163, 168
ejaculation   58–60, 63, 71, 91, 151
electrolytes   39, 92
electron microscopy   8
– probe analysis   43, 44, 48
electrophoresis   36, 37
embryo   1, 2, 7, 14, 15, 19, 20, 22, 26, 30, 33, 34, 36–39, 42–46, 48, 57, 66, 102, 109, 112–114, 117, 120–122, 132, 139, 141, 143, 145, 168–171, 174, 176
– culture   48, 114, 115, 122, 149, 169
– transplantation   168–171, 176
– transport   127–136, 143, 145
embryonic death   66, 73, 74, 101, 102, 118, 120, 127, 141, 167
– development   7, 14, 30, 33, 37, 41–43, 48, 84, 100, 101, 110, 113, 114, 119, 120, 127, 145, 147, 149
– disc   121
– nutrition   84, 113, 114
– origin   14
– programming   20, 120
– signals   26, 110, 115
– testes   14
– transplantation   9, 30, 113, 118–120, 127, 136, 140, 143, 145, 149
endocrine conditions   7, 46
– disturbance   14, 127, 135
– imbalance   127, 135
– information   19, 46

− status   19
endoderm   112, 122
endometrium   47, 120, 147, 149, 168
endosalpinx   30, 31, 38, 48, 58
energy substrates   41, 113−115
enzyme inhibitors   92
enzymes   38, 84, 89, 92, 93
epididymal spermatozoa   162−164, 167
− storage   72, 91
epididymis   16, 91, 92, 95, 140
epithelium   14, 61, 109, 113, 117, 143, 144
equatorial segment   93, 94, 96, 97
equids   130
Estes' operation   8, 145−147
exocrine secretion   46
exocytosis   99

Fabricius   5
Fallopius   1, 4
female tract   14
ferret   110
fertility   53, 139
fertilizable lifespan   68
fertilization   7, 9, 30, 36, 38, 41, 42, 57,
   59, 60, 62−65, 67, 70, 72, 81, 84, 85, 88,
   91, 93, 95, 102, 109, 112, 114, 116, 118,
   145, 147−149, 160−163
− abnormalities   64−66, 101, 151
− in vitro   30, 33, 85, 110, 115, 122, 140,
   143, 149, 156, 160−165, 168, 169, 172,
   174, 183
fertilizin   84, 85, 161
fertilizing potential   62
fimbria   12, 13, 17, 18, 23, 54−57, 69, 70, 171
flagellar beat   93
fluid collection   31−34, 41, 43
− composition   34, 36−42, 44−48, 67, 73
− flow   33−36, 57, 58, 67, 132, 144
− formation   31−34
− volume   31−34, 37, 43, 44, 57
foetus   1, 118, 121
follicle   4, 7, 26, 41, 47, 53−57, 74, 83, 99,
   100, 148, 165, 167, 172
− selection   74, 167
follicular cells   56, 81−84, 102, 113, 163
− fluid   46, 48, 54, 56, 57, 67, 69, 92,
   163−165, 167
− hormones   90
− phase   20, 24, 26, 36, 37, 43, 165
freeze-fracture   93, 94, 97

Galen   2
gamete lifespans   72−74
gametes   4, 8, 14, 22, 24, 30, 34, 37−39,
   41−45, 53, 57, 62, 72−74, 81, 85, 86,
   90, 96, 102, 139, 151, 153, 160, 162, 163,
   166, 171, 172

gene therapy   176
genes   102, 121, 176
genetic abnormalities   72
− engineering   176
− errors   102
− maleness   14
− screening   173
genital duct   15
− tract   14
genome   71, 109, 115, 116, 121
− reconstitution   177
germ cells   140
GIFT   171, 172
globulin   37
glycogen   41
glycoprotein   36−38, 84, 85, 99
goat   8, 58, 172, 175
Golgi apparatus   93, 98
gonad   12, 15, 16, 18, 19
gonadal hormones   26, 30, 34, 38, 39, 132,
   143
gonadotrophin surge   54, 82, 92, 97, 98,
   100, 112, 165−167
− treatment   39, 45, 74, 148, 165, 166
gonadotrophins   45, 82, 100, 148, 165
Graaf, de   1, 4
Graafian follicle   4, 7, 17, 36, 46, 47, 54,
   55, 67, 68, 73, 81−83, 86, 90, 91, 99,
   100, 118, 131, 167, 174
granulosa cells   41, 46, 54, 56, 82−84, 86,
   113, 167
guinea pig   18, 20, 24, 67, 70, 110, 134,
   145, 161, 162, 171
gynaecology   2

hamster   46, 58, 86, 90, 92, 93, 100, 110,
   118, 134, 147, 148, 162, 163, 165, 173
haploid gene expression   71
hares   73
Harvey, William   4, 5
hatching   112, 120
Herophilus   2
Hindu medical writings   1
hormone concentrations   46, 47
hormones   45−48, 68
Horne, van   4
horses   13, 18, 24, 54, 58, 60, 62, 63, 68,
   73, 89, 92, 100, 103, 110, 120, 129
human   8, 9, 15, 18−20, 22, 23, 26, 36, 38,
   40, 41, 54, 55, 57, 60, 93, 110, 112, 114,
   115, 118, 119, 122, 131, 132, 136, 144,
   151, 162, 163, 165, 169, 170, 173, 177,
   183, 184
− chorionic gonadotrophin   142, 166, 170,
   171, 172
H-Y antigen   14
hyaluronidase   83, 84, 93

hybridization 172
hydrosalpinx 43, 140
hyperactivation 39, 68, 89, 91, 92, 95, 96
hypogastric nerve 23
– plexus 22, 23
hysterectomy 144, 145

immotile cilia syndrome 26, 58, 132
immunization 48, 134, 165
immunological masking 111
implantation 112, 115, 116, 118, 134, 141, 144
in vitro fertilization 9
indomethacin 134
induced ovulation 4
infertility 68, 139, 143, 160, 161, 165, 166, 171–173
inflammatory conditions 140, 143
infundibulum 12, 17, 20, 22, 26, 43, 44, 55, 69
inhibin 165
inner cell mass 121, 122
innervation 14, 22, 131, 135
insufflation test 8
intercellular bridges 111
intersex animals 14, 16, 140
intersexuality 140
intra-mural portion 12, 13, 18, 20, 22, 24, 47, 68, 115, 130, 141, 143, 144
intra-peritoneal insemination 70, 71, 165, 171, 172
intra-uterine device 143
– insemination 165, 172
ionic composition 39
isthmus 12, 13, 19–24, 26, 34, 35, 37, 41, 43–47, 59, 61–64, 66–71, 103, 112, 128–131, 133, 136, 140, 149, 151, 153, 171

Kartagener's syndrome 58
karyotype 102, 173
Kreb's cycle 37
kymographic technique 7

lactate 113, 114, 164
laparoscopy puncture 83, 152, 156, 166, 167
laparoscopy 140, 142, 152, 156, 167, 171
laparotomy 17, 18, 36, 66, 100, 156
lectins 85
leucocytes 102
Leydig cells 140
ligation 43, 120, 145, 148, 152
lipid domains 92
lumbar nodes 19
luminal fluid 8, 36–48
luteal function 116, 167

– phase 20, 24, 31, 32, 39, 41, 43, 167
luteolysin 144, 145
luteolytic factors 19, 144
lymph flow 20, 45
lymphatic system 14, 19, 20, 24, 31, 69
– vessels 19–21, 25
lymphocytes 116, 117
lysosome 93

male tract 14
marmosets 119, 134
marsupials 56
maternal recognition 118
mating 4, 53, 58, 59, 62, 67–71, 73, 89, 92, 102, 150
meiotic division 97, 174
– spindle 74, 98, 101
membrane fusion 99, 101
– vesiculation 83, 84, 93, 99
menstrual cycle 18, 24, 26, 31–33, 39, 43, 47, 134
– fluid 2, 22
menstruation 19, 37
mesenteries 7, 13, 18, 55, 57, 130, 151, 171
mesometrium 19, 130
mesonephric ducts 14, 15
mesosalpinx 13, 19, 130, 156
metabolic activity 92, 109, 112, 113
– clearance 47
microdrops 164, 165
microenvironment 43–45, 48, 67, 113, 114
microflow 34, 35
microinjections 66, 73, 122, 133, 175
microsurgery 8, 9, 43, 44, 111, 112, 135, 151, 153, 154, 156
microtubules 74, 98, 101
microvasculature 19
microvilli 26, 96, 121, 122
mink 110
mitochondria 122
mitosis 99, 120, 168, 176
molecular embryology 121
monkey 8, 24, 26, 36, 57, 67
monoclonal antibodies 48
morphogenetic changes 14
morphology 12, 13, 71
morula 110, 116, 120, 121, 169
mosaic membrane 97
mouse 8, 12, 38, 41, 44, 45, 48, 58, 67, 71–73, 92, 110, 112–115, 117, 121, 122, 134, 145, 162, 163, 175
mucification 54
mucin layer 7, 26, 38, 112, 129
mucoproteins 44
mucosa 19, 21, 23, 24, 31, 34, 41, 47, 64, 66, 128, 130, 132, 133, 140, 156

mucosal folds   12
mucus   26, 59, 60, 130
Müllerian duct   14, 15, 140
muscle layers   19–22, 64, 66, 128
– tone   48
muscular contractions   8, 48, 57, 59, 61, 66, 130, 134, 170
musculature   14, 20, 57, 61, 130, 131, 141, 143
myometrial contractions   61, 143
myosalpingeal contractions   48, 71, 83
myosalpinx   20, 23, 31, 57, 68, 83, 115, 128, 130–132, 135, 143, 151, 183

negative feedback   30
neuro-anatomy   22
neuropharmacology   8
non-disjunction   101
noradrenaline   22
nuclear membrane   97
– structures   111
– transplantation   121, 175, 176
nutritional substrates   113, 114

oedema   24, 31, 34, 64, 66, 130, 132, 133
oestradiol   31, 32, 34, 46, 131, 167
oestrogens   18, 24, 31, 33, 36, 37, 39, 43, 57, 73, 127, 130, 131, 133–135, 143
oestrone   46
oestrous cycle   7, 18, 24, 26, 30–32, 34, 36, 39, 113, 118, 145
oestrus   31, 32, 34, 36, 39, 41, 43, 53, 61, 68, 73, 103, 130, 132
oligospermy   168, 175
oocyte   4, 39, 48, 53–58, 74, 81, 82, 84, 97, 100, 111–113, 115, 145, 148, 152, 156, 160, 162, 165, 166, 173, 174
– aspiration   83, 148, 162, 165–167, 174
– maturation   113, 160, 163, 167, 174
– metabolism   57, 113
osmolality   43, 163
ovarian artery   18, 20, 47, 69
– bursa   12, 33, 36, 54, 102
– cycle   13, 30, 36
– hilus   20
– hormones   24, 26, 43, 55, 57, 68, 73, 81, 91, 113, 127, 128, 133, 134, 168
– hyperstimulation   166, 167
– plexus   20, 22, 23
– response   19
– tissue   16
– transplants   8, 145, 146
– vein   47, 69
ovariectomy   31, 32, 37, 39, 43, 127
ovaries   1, 2, 4, 7, 15, 17, 19, 20, 53, 54, 57, 69, 100, 115, 118, 127, 140, 145, 146
ovotestis   14, 16, 140

ovulation   4, 18, 24, 26, 30–33, 35–37, 39–41, 43, 44, 46, 47, 53–58, 61, 62, 67–70, 82, 84, 86, 89–92, 97, 99, 100, 103, 110, 112, 113, 118, 127–130, 134, 145, 162, 167, 171, 174
ovum   4
– factor   117, 118
oxygen tension   43
oxytocin   45, 47, 48, 61

para-aortic nodes   19
paramesonephric ducts   14
parasympathetic system   23
parthenogenesis   161
pathology   140
penis   58
peptide hormones   45, 46, 48, 54, 69, 84
peptidergic nerves   23
peristalsis   130, 135, 136
peritoneal cavity   12, 33, 58, 66, 70, 102, 144
– fluid   36, 46, 62, 103
perivitelline space   96, 99
pH   43, 113, 167
phagocytosis   62, 69, 71, 97, 102, 103, 122
pharmacological techniques   7, 135
pharmacology   8
physiological sphincter   22, 131
pig   7, 8, 12, 13, 16–18, 20, 21, 24, 25, 32, 35, 39, 43, 45, 54, 57, 58, 60–71, 73, 74, 82, 84, 89, 90, 92, 99, 100, 110, 112, 117, 120, 128–132, 148–150, 162, 163, 169, 171
placenta   121, 142
plasma membrane   37, 61, 70, 85, 92, 93, 96–99, 101, 173
platelet-activating factor   118, 119
polymorphs   69, 102
polyploids   101
polyploid processes   25
polyspermic fertilization   64–66, 73, 74, 84, 96, 99, 101, 150, 151, 168, 173
Pomeroy technique   152
Pontamine blue reaction   115
post-nuclear cap   97
post-ganglionic fibres   23
potassium ions   67
pouch of Douglas   103, 151, 172
pre-ganglionic fibres   23
pregnancy   7, 18, 19, 26, 30, 31, 43, 73, 115, 118, 119, 135, 141, 142, 145, 149, 153, 156, 169, 170
prenatal loss   74
primary oocyte   112, 174
primates   12, 24, 26, 30, 32, 39, 41, 53, 54, 58, 59, 68, 89, 109, 110, 115, 118, 120, 122, 127, 128, 130, 134, 135, 141, 144, 149, 163, 169, 184

proacrosin 93
progesterone 24, 31, 32, 34, 37, 41, 43, 46, 48, 57, 66, 73, 109, 116, 120, 127, 131, 133–135, 144, 167
prolactin 45, 47
pronase 112
pronuclear membrane 97
pronucleus 64, 97, 99, 101, 150, 161, 168, 174, 176
prostaglandins 8, 45–48, 54, 60, 61, 69, 90, 131, 134
prostaglandin E 131
– $F_{2\alpha}$ 19, 47, 131, 134, 144
protein content 36, 37, 42
– secretion 14
proteolytic enzymes 84, 93
pseudopregnancy 37, 46, 120
psychosomatic disturbancy 135
puberty 13
pulmonary circulation 47
pulsatile treatment 166, 167
pyruvate 41, 42, 67, 84, 113, 114

rabbit 4, 8, 12, 24, 26, 31, 33, 37–39, 41–43, 46, 54, 56–59, 61, 62, 67, 70, 71, 82, 88, 90, 92, 93, 99, 100, 110, 112, 116, 120, 128, 129, 131, 132, 134, 145, 147, 149–151, 161, 162, 171, 172
rat 12, 24, 26, 45, 54, 58, 63, 67, 88, 90, 110, 115, 134, 162, 172
reconstructive surgery 68, 73
relaxin 45, 47, 48
respiration 39, 41
rhesus monkey 18, 24, 31, 32, 37, 38, 41, 43, 46, 110, 115, 130, 134, 149, 169
ritodrine 135
rodents 13, 58, 62, 82, 99, 102, 109, 114, 115, 120, 121, 162, 163, 169
rosette inhibition test 116–118
ruminants 58, 59, 110

salpingectomy 145
salpingitis 140, 143
scanning electron microscopy 8, 26, 35, 86, 93, 96, 115, 117, 132, 133
second polar body 97, 98
secretion 18, 30, 31, 36–38, 41, 44, 48, 57, 62, 168
secretory activity 26, 37
– cells 8, 14, 24, 26
– granules 26, 37
– pressure 31
semen 2, 4, 24, 25, 58–62, 64, 69, 89
seminal plasma 38, 60, 61, 63, 69, 72, 91, 92, 163, 168, 171–173
serosa 19–21, 66
serous coat 19

Sertoli cell 14
– cells 140
sex chromatin 176
– selection 176
sheep 8, 12, 13, 18, 24, 26, 31–34, 36, 38, 39, 42, 45, 46, 54, 56, 58, 60–62, 67–69, 73, 82, 84, 90, 93, 100, 103, 109, 110, 112, 117, 120, 129, 132, 162, 163, 169, 172, 176
site of fertilization 13, 26, 53, 57, 58, 61–64, 67–72, 83, 85, 89, 91, 100–102, 128, 132, 150, 151
Soranus 2
spasm 61, 144
sperm agglutination 72, 84, 85, 164
– arrest 67–70
– attachment 86, 88
– binding 85, 88, 172
– concentration 60, 168
– distribution 67–69, 90
– flagellum 67, 88, 96
– gradient 59, 61, 63, 64, 66, 67, 71, 151
– head 65, 93–97, 161, 173–175
– karyotype 102
– loss 102
– maturation 70, 91, 95
– metabolism 60, 92
– mitochondria 88, 95
– motility 59, 61, 63, 67, 91, 92, 168
– penetration 53, 59, 65, 72, 85, 88, 95, 97, 99–101, 113, 148, 160
– redundancy 71
– release 67–69
– reservoir 59, 67–70, 112, 128, 171
– selection 69, 71, 72
– storage 22, 59, 67–70, 128
– survival 70, 73
– suspension 65, 66, 70, 85, 92, 149, 150, 161, 163, 164, 167, 168, 171
– transport 22, 24, 57–59, 61–63, 66–69, 84, 91, 128, 132
– viability 70–73
spermatogenesis 72, 140
spermatozoa 7, 34, 36, 38, 39, 41, 43, 48, 53, 57–73, 81–85, 100, 112, 132, 147, 149, 150, 161–163, 168, 172
sphincter 12, 22
spontaneous ovulators 67, 99, 101
sterility 139
sterilization 37, 153
steroid hormones 24, 30, 34, 38, 39, 43, 45, 46, 48, 54, 55, 57, 69, 73, 74, 84, 109, 113, 127, 128, 131–134, 143, 165, 168
subserous coat 19–21
superovulation 100, 160
surgical anastomosis 66

– insemination 65, 71, 92, 112, 171
– studies 8
sympathetic fibres 22, 23, 54
synchronization 30, 168, 169
syncytiotrophoblast 136, 144
synergism 92
syngamy 64

temperature 67
testes 2, 14
testosterone 46
thecal layers 54
thrombocytopenia 117, 118
totipotency 111, 120, 175
transection 70
translocation 14
transperitoneal passage 70
transudation 18, 31, 36, 37, 44–46
triplets 111
triploids 101
trophectoderm 112, 121, 122
trophoblast 103, 121, 122, 142, 145, 176
tubal anastomosis 8, 66, 150, 152–156
– anomalies 140
– arcade 18, 19
– epithelium 8, 14, 24, 26, 111, 114, 115
– fluid 8, 18, 30–33, 36–48, 57, 67, 83,
  91, 109, 112–114, 120, 130, 164
– isthmus 8, 130, 131, 149, 150
– occlusion 143, 145, 146, 148, 153
– ostium 54, 56, 57, 70, 171
– patency 8, 66, 67, 69, 103, 128, 132,
  143, 153
– pregnancy 135, 136, 141–144
– resection 8, 66, 149–152
– spasm 135, 144
– surgery 144, 149–156
– transection 68, 69
tube locking 133, 134
twins 111

ultrasonic scanning 142, 167, 171, 172
ungulate 13, 26, 112
uterine artery 18
– blood 2
– contractions 60
– fluid 24, 36, 47, 48, 61, 168
– flushings 59

– fundus 20
– horns 2, 20, 59, 61, 70, 101
– lumen 4, 33, 69, 70, 110
– secretions 109, 115, 120, 144
– wall 19, 20
uteroglobin 37
utero-ovarian vein 19
utero-tubal junction 12, 13, 20, 21, 24, 25,
  33, 34, 36, 44, 59, 61–64, 68, 71, 72, 91,
  140, 144, 150, 151
uterus 1, 2, 4, 7, 8, 15, 18, 20, 21, 30, 33,
  34, 44–46, 58–62, 71, 92, 100–103,
  109, 110, 115, 119, 127–129, 132–134,
  140, 141, 143, 147–149, 160, 161, 169

vagal fibres 23
vagina 7, 58–60, 63, 170
vascular anatomy 18, 69
– bed 30, 36, 45
– perfusion 42
– system 14, 18, 69
vasculature 17, 18
venous plexus 19
Vesalius, Andreas 1–4
vesiculation reaction 93
Vinci, Leonardo da 2
vitelline membrane 99, 101
– surface 97, 101
vitellus 72, 95–97, 99, 101, 150

whiplash motility 95
Wolffian duct 14, 15, 16, 140
women 12, 18–20, 24, 26, 32, 33, 36,
  38–41, 47, 53, 58–60, 68, 69, 73, 74,
  82, 100, 103, 110, 118, 119, 130, 132,
  134, 135, 139, 149, 152, 184

Y-chromosome 14
yolk 112

zona pellucida 54, 82, 85–88, 93, 95, 96,
  99, 100, 102, 103, 111, 112, 115, 117,
  120, 121, 129, 147, 172, 174
zona-free egg 102, 112, 173, 174
zygote 8, 30, 41, 44, 45, 113–115, 118,
  136, 166, 171, 175
zygotin 117, 118
zymogen 93